Insect Conservation in Australia: Why and How

Tim R. New

Insect Conservation in Australia: Why and How

Springer

Tim R. New
Department of Environment and Genetics
La Trobe University
Bundoora, VIC, Australia

ISBN 978-3-031-66630-8 ISBN 978-3-031-66631-5 (eBook)
https://doi.org/10.1007/978-3-031-66631-5

This Springer imprint is published by the registered company Springer Nature Switzerland AG
The registered company address is: Gewerbestrasse 11, 6330 Cham, Switzerland

Preface

Over nearly 50 years of trying to promote interest in insect conservation in Australia, the most frequent responses to my pleas for understanding have been questions along the lines of 'why bother?' and 'why do insects matter?', sometimes accompanied by 'how do we do it, anyway?'. So widespread and entrenched are feelings that insects are dangerous or harmful, 'pests', a 'nuisance' or even 'enemies' to human welfare and should be suppressed, that my responses to the first questions—advocating insect values—have commonly been received with cynicism and disbelief. Most people do not love—or even approve of—insects! There are, of course, exceptions. A few insect groups, most notably butterflies and dragonflies, are sufficiently popular to be considered 'iconic'. The paramount roles of butterflies in global insect conservation activities, for example, are possible only because many people indeed like them, and many insects have places in folklore or human culture—as examples, dragonflies are embedded in the traditional culture of Japan, and the ball-rolling activities of some dung beetles in that of ancient Egypt, where 'scarab' amulets became commonplace. Fireflies have widespread popularity for tourism. In short, insects enrich many aspects of human life. Such approval from aesthetic and cultural values links intrinsically with acceptance of worth and, in turn, that these insects merit conservation. Some insects are indeed accepted as 'good'!

More pragmatically, provision of honey or silk renders their producers—particular bees or moths—of economic and social value, together with a variety of insects harvested for food, trade, or medical purposes. Butterflies, dragonflies, and bees as the most 'popular' or 'iconic' insects comprise a very small proportion of insect life. For most insect orders, such positive or materialistic attributes that give them tangible 'worth' are more elusive, and any conservation sympathy may more embody the basic ethical sentiment that no species should be allowed to become extinct from human cupidity or—more pragmatically—that insects can be ecologically important and in some way benefit human wellbeing. However, these principles rarely compensate fully for insect conservation being considered widely to be low priority in the spectrum of conservation urgency or, even, the realm of 'cranks', and belief that insects should really be 'squashed'. Redressing this perception is recognised widely as a critical need, and numerous essays extolling the importance of insects embrace the above and other ways in which humanity may benefit from insect life, and how this may increase in the future (Scudder 2009). However, far more has been written on how to 'control' insects than on how to conserve them, with suppression of insect disease vectors and economically damaging pests both major facets of protecting global human health and food supply.

Much has indeed changed in Australia since I expressed concern about 40 years ago (New 1984), but despite strong further advocacy for insect conservation (such as by Yen and Butcher 1997) and continuing activities of both professional and non-professional groups, much still remains to be accomplished to justify the importance and needs of insects and establish these on the wider national conservation agenda. The problems are, of course, far more widespread. As Goulson (2019) presciently commented 'Ecologists and entomologists should be deeply concerned that we have done such a poor job of explaining the vital importance of insects to the general public'. Public and political sympathy is an urgent and vital element in insect conservation and mitigating further declines, and attempts to encourage this through education and 'citizen science' continue to proliferate, and to diversify from the awareness initiated through concerns for butterflies and bees. Perhaps the impetus to encourage people to 'like insects' is greatest in urban and near-urban areas, where most people live and contacts with insects are consequently most common (Hunter and Hunter 2008), together with fears for their real or imagined adverse impacts.

The above questions of insect worth and need to sustain them are nevertheless sincerely posed and seek the evidence to counter the sentiments expressed. They also reflect that the great majority of people charged with important decisions over managing the natural world lack any formal entomological background to guide their directions and priorities. Not unreasonably, they then rely on the greater collective experience and lessons available on conserving larger vertebrates and some vascular plants, rather than incorporating the enormous (and indeed daunting) ecological and taxonomic diversity of insect life—so that insects and other invertebrates are commonly relegated to being 'passengers' in conservation programmes that target more popular taxa. Interest in their specific conservation has lagged considerably behind that for mammals and birds, for example. As Holwell and Andrew (2015) put it, although most people may be familiar with a Platypus or Kiwi 'many may not appreciate that the insect fauna of Australia and New Zealand is equally unique and far more diverse'.

This lag is not for want of importance. Literature on the ecological roles of insects and other invertebrates is vast, continues to burgeon, and in various ways illustrates that insects are functionally vital in many ecosystems.

The purpose of this essay is to display this importance to non-entomologists, to show why insects matter, how the fate of humanity may depend on insect activity and persistence—and, so, why the conservation of insects is needed in many terrestrial and freshwater ecosystems and links strongly with the wellbeing of people. Wilson's (1987) widely quoted axiom to the effect that 'if humans became extinct, the world would carry on as usual but if insects became extinct the world would change drastically' has been both ridiculed and applauded—and many ecologists now consider it a very reasonable encapsulation of our prospects. Similar sentiments have been advanced repeatedly. The principle resonates strongly and is difficult to refute across all ecosystems that insects frequent, many of which they dominate in terms of variety, biomass, and influence, and that may be changed severely by human activities. They are also the arenas in which insect conservation is needed and pursued most strenuously in many parts of the world, as the impacts of those anthropogenic disturbances are realised, but still proliferate and intensify.

My main focus here, the Australian insect fauna, is no exception but is still inadequately documented so adding a dimension that many people find difficult to accept—that conservation must embrace large numbers of undescribed, even undiscovered species of insects and other invertebrates, and that local diversity and peculiarities can often be evaluated only very imprecisely. The adage 'out of sight, out of mind' applies to most insect life. Not only do we not know how many insects occur in particular ecosystems and what they are, but we also cannot evaluate properly what most of them do and what the consequences of their loss may be.

This book is a general overview of how the importance of insects to humanity and in the natural world may be assessed and communicated to those whose decisions and actions may determine their fate, and how much-needed conservation may be pursued constructively. Most content is not new and will be familiar to entomology-aware readers—but the importance of insect life deserves frequent endorsement and needs strong evidential and philosophical defence to penetrate all sectors of our community, ranging from young people to leading politicians and government agencies, in order to gain more than casual attention to its wellbeing. This vision, of course, extends well beyond insects, but this principal component of natural biodiversity cannot continue to be ignored or diminished in importance. This has become increasingly apparent over recent decades, with numerous reports of insect declines and concerns of the wider impacts of these driving calls for greater understanding and conservation action. For people to appreciate the world as we currently know it, and for it to function, we need insects—without them, likely rapid consequences include losses of insectivorous birds and many flowering plants, trends already well documented in the northern hemisphere and that may cascade to far wider ecological disruption.

The book comprises two main parts, together encompassing the 'why' and 'how' of my title. The first part details the problems of studying and enumerating insect faunas, initially through their massive taxonomic and ecological diversity, often difficult to evaluate (Chap. 1) and with vociferous concerns over recent declines (Chap. 2). Many of the reasons for concern flow from anthropogenic activities and their numerous and varied threats to insects (Chap. 3), some with severe and unpredicted consequences (Chap. 4). The second section discusses human perceptions of insects and their worth (Chap. 5), and the major reasons for needing insect conservation (Chap. 6), before moving to practical aspects of how insect conservation may be promoted and attempted (Chap. 7) and interest and participation expanded through the human community (Chap. 8). I conclude with a short suggested agenda for the future of insect conservation in Australia (Chap. 9) to indicate how the many complementary strands of practical insect conservation discussed in earlier chapters, and relevant elsewhere in the world, may be focused and integrated. The book is intended primarily for use by non-entomologists, and I have included some basic information and references as well as citations for the more technical points discussed, using literature I have seen up to early 2024. For the same reason, I have added an illustrated Appendix to help readers recognise the major orders of insects that have significant roles in conservation at present and their major importance to humanity. Reflecting the Publisher's need for each chapter to be read independently, some themes recur in different contexts in different chapters: page cross-references facilitate integration.

Bundoora, VIC, Australia Tim R. New

References

Goulson D (2019) The insect apocalypse, and why it matters. Curr Biol 9:R967–R971

Holwell GI, Andrew NR (2015) Protecting the small majority; insect conservation in Australia and New Zealand. In: Stow A, Maclean N, Holwell GI (eds) Austral Ark: the state of wildlife in Australia and New Zealand. Cambridge University Press, Cambridge, pp 278–297

Hunter MR, Hunter MD (2008) Designing for conservation of insects in the built environment. Insect Conserv Divers 1:189–196. https://doi.org/10.1111/j.1752-4598.2008.00024.x

New TR (1984) Insect conservation, an Australian perspective. W. Junk, The Hague

Scudder GGE (2009) The importance of insects. In: Foottit RG, Adler PH (eds) Insect biodiversity, science and society (first edition also 2nd edn, 2017, pp 9–43). Blackwell Publishing, Chichester, pp 7–32

Wilson EO (1987) The little things that run the world (the importance and conservation of invertebrates). Conserv Biol 1:344–346

Yen A, Butcher RJ (1997) An overview of the conservation of non-marine invertebrates in Australia. Environment Australia, Canberra

Acknowledgements

I thank the following for granting permission to reproduce illustrations or tables from publications to which they hold copyright; some others are used under the provisions of Creative Commons licences, as noted individually in headings or legends: Elsevier, John Wiley and Sons, Oxford University Press, Springer Nature. Every effort has been made to obtain these permissions and the Publishers would appreciate advice of any inadvertent omissions or corrections that should be included in any future imprints or editions of this book.

I am very grateful for the continuing interest and support of Dr Lars Koerner who, as Editor, has provided much advice and practical help. I also thank my Project Manager, Mr Srinivasan Manavalan. The careful attention to detail from Cineha D in preparing the book for publication is very much appreciated.

Contents

Chapter 1
Insect Diversity, Distribution, and Abundance

1.1 Introduction

Insects are the predominant animals in most terrestrial and freshwater ecosystems across the world, and are heralded widely as the animal group containing the most different life forms or 'species'. They are thus 'megadiverse' and, although rivalled in the richness of species by mites (Acarina) and nematode worms (Nematoda), are far more visible to people, with a number of their key life forms familiar and distinguished and recognised easily (Appendix). Discussions continue over the relative species richness of these three massive invertebrate groups. However, many authorities accept readily that insects are both the most diverse and the most ecologically widespread of all animal taxa, and have flourished over more than 400 million years of Earth's history, with their first appearance in the fossil record about 420 million years ago. Over that long evolutionary history, vast numbers of species have become extinct, but the fossil record confirms that many representatives of the lineages and insect orders found today manifest a long period of evolution and diversification. Yet, despite their recent predominance, we still have only very imprecise estimates of the number of insect species living in many parts of the world. Taxonomic diversity also reflects ecological and functional diversity. Insects participate in vast numbers of interactions—with each other, plants and/or other animals—in almost all terrestrial and freshwater ecosystems and, in many cases, are essential to sustaining these ecological intricacies.

Perhaps only in western Europe and North America can reasonably definitive insect inventories and regional species lists be derived—but even in the best-known of all insect faunas, that of Great Britain, novel species continue to attract notice in most years, building on a legacy of more than two centuries of concentrated field study, collection and description, and documentation of their biology and distribution. For example, and perhaps reflecting more widespread outcomes of changing climate conditions, a European barklouse (Psocoptera) and a grasshopper

T. R. New, *Insect Conservation in Australia: Why and How*, https://doi.org/10.1007/978-3-031-66631-5_1

(Orthoptera) were reported as new to Britain in the July–September 2023 issue of the Entomologist's Monthly Magazine. More surprisingly, a tiny moth found recently in west London was not only undescribed but proved to be a species from Western Australia! An individual of the same species collected in Western Australia in 1886 had remained undescribed in the collections of the Natural History Museum in London; comparisons using subtle DNA barcode investigations confirmed that these were indeed the same species (Sterling et al. 2023), and the moth (now described as *Tachystola mulliganae*, Oecophoridae, Australia's largest moth family) appears to be established in England, possibly having been imported in plant material. Britain now has around 24,000 insect species (the number varying with different accounts: Duffus and Morimoto 2022, cited 24,043 species for the United Kingdom and Ireland), many of which have been studied in some detail over a considerable period so that their distribution, biology, conservation status, and needs can be appraised reliably, and changes detected and monitored. And, as above, additional species records can be detected with reasonable confidence, flowing from good knowledge of the resident insect fauna, so that 'oddities' are both sought and recognised. Even in that fauna, however, some insect orders are still incompletely known and are studied by few people. As the best-documented insect component of any regional fauna, the extensive literature on British butterflies and larger moths provides an enviable and unique foundation for assessing changes in incidence, abundance, and distributions, so that changes in both space and time can be appraised with reasonable confidence. With the continuing monitoring schemes possible with the participation of large numbers of enthusiasts and a collective large number of sampling sites throughout the region, it is often feasible to claim that 'species A' has recently changed in range and in seasonal activity. Such trends, discussed in Chap. 2, cannot be detected easily for many lesser-studied taxa even in that restricted fauna. For the British insect fauna, Morris and Welch (2023a) commented that 'we still don't know what occurs where', noting that very few inventories exist for significant areas such as designated National Nature Reserves and Sites of Special Scientific Interest, and that the expertise to redress this is manifestly inadequate.

Despite the commonalities of these gaps with Australia, there are also marked contrasts. The first point is that Australia's insect fauna is vastly greater—perhaps with at least ten times the number of species found in Britain—and far less completely documented. The total of British insect species is approached by Australia's moths alone, but around half of Australia's 20,000 or so moth species are yet to be diagnosed and named (and one recent figure suggested a further 10,000 species may be present). A second point is that Australia has far fewer entomological observers and enthusiasts to deal with this massive and daunting richness, and has also had a far shorter period of natural history exploration, with that dominated by non-insect interests. Australia's land area is more than 30 times that of the United Kingdom and spans tropical to cool temperate regimes and wet to arid environments to give an enormous variety of local environments and vegetation systems that can each support localised insect life. Much of the country is remote, and its insect fauna underexplored—other than, perhaps by sporadic 'snapshots' from brief survey visits—and remains difficult to explore thoroughly: this again is a marked contrast to the detailed information on insect distributions available across most of Britain. Not surprisingly, the most complete knowledge of Australian insects is from the eastern/southern coastal and near-coastal regions that harbour most post-European human settlement, and from the far southwest corner. Even there, in the least disturbed areas close to capital cities, documentation of insect orders is uneven and incomplete. Butterflies are the most completely assessed taxonomic group, reflecting hobbyist interests, but there is no equivalent to the extensive recording schemes pursued in Britain under the aegis of the organisation 'Butterfly Conservation', now having coordinated such initiatives for more than 40 years. In the then most intensively investigated State of Australia, Victoria, a pioneering attempt by the Entomological Society of Victoria to map the state's butterflies on a grid scale (using 10′ × 10′ areas, so a very coarse scale, each unit with a variety of resources and habitats) showed that about half of those units had no butterflies recorded from them by 1983, and the map produced by the Society (Fig. 1.1) showed the pattern of major highways and traditional collecting sites rather than reflecting any true pattern of distributions and species incidence. Likewise, the British light-trapping programme for

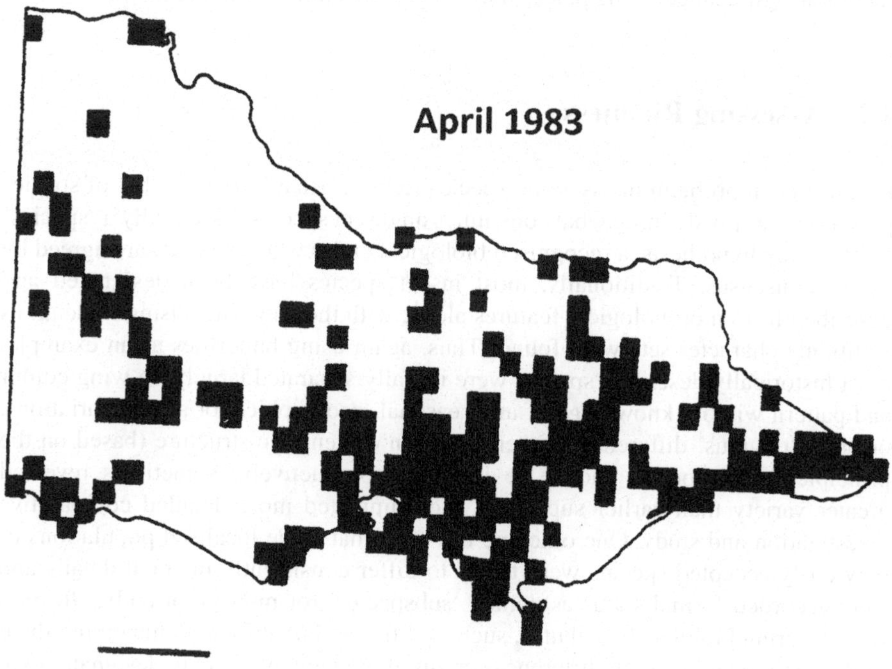

Fig. 1.1 Early records of butterflies in Victoria to illustrate the incomplete historical base for assessing distribution and conservation status. The distribution of 10 × 10 min (ca 18 × 15 Km) areas from which any butterflies had been recorded by April 1983 under the Entomological Society of Victoria's ENTRECS scheme; black areas have butterfly records, open areas had no detected records. Scale line equals 100 Km. Note the concentration of collecting around Melbourne (central south of map) and the importance of the Western Highway from Melbourne towards the northwest (after New 1984)

moths (p. 26) is far more complete than any data currently available in Australia, despite recently determined efforts by a number of dedicated enthusiasts. A shortage of entomologists is a severe impediment to discovering more of Australia's insect life, and to assessing conservation needs. A consequence of high insect richness, imperfect knowledge, and limited naturalist attention is that comprehensive illustrated 'field guides' or 'handbooks' to identify most insect groups do not exist, so that difficulty of gaining a foothold of information on many taxa can be a deterrent to further interest. The important exceptions span many of the taxa of larger insects now of most concern in conservation [Odonata (dragonflies and damselflies), Orthoptera (and allies such as Phasmatodea), Lepidoptera (butterflies and some larger moths)] but much of the literature facilitating species identifications remains in the specialist domain of formal scientific papers.

In short, enumerating and fully characterising Australia's insect fauna is a 'work in progress', but a total of more than 200,000 species has been suggested (Yeates et al. 2003), and even this may be a substantial underestimate. There is no shortage of notable species, many apparently locally endemic, restricted to particular habitats and representing isolated phylogenetic lineages—and, so, of potential or actual conservation significance as the places they inhabit succumb to human intervention.

1.2 Assessing Richness

One universal problem in assessing species richness (that is, the number of species present) is simply defining what constitutes an insect species. Essentially a 'species' is a working hypothesis, a recognised biological entity whose bounds are agreed by expert consensus. Traditionally, most insect species have been designated and described from morphological features alone, with the latter increasing in detail as additional character sets were found. Thus, again using butterflies as an example, most historically described species were initially separated largely on wing colour and pattern without knowledge of any seasonal, geographical or sexual variations, or other 'obvious' differences. Later inclusion of genitalic structure (based on the principle that a species should be isolated reproductively) sometimes revealed greater variety than earlier supposed, and stimulated more detailed comparative investigation and study. One outcome has been that some localised populations of previously accepted species were found to differ consistently in small details and were accorded formal status as named 'subspecies' (or more informally, 'forms') with differing biological attributes, such as different larval food plants giving them further individuality. Such differences are used by many workers to designate 'evolutionarily significant units' (ESUs) of parallel conservation worth to 'true species' because they imply additional biological and evolutionary variety. However, others doubt whether the formal status of 'subspecies' should be given value in conservation (Braby et al. 2012). Numbers of ESUs continue to increase as relevant investigations into species complexes continue. Increasing use of molecular methods—mostly DNA barcoding—adds further dimensions to assessing insect biodiversity, with many previously cryptic species and 'species complexes' containing many consistently distinct life forms.

Some classic and influential cases are noted here to illustrate the nature of such revelations. Perhaps the best-known example of this then innovative technique is of the neotropical skipper butterfly known as *Astraptes fulgerator*, which was expanded to ten distinct species on consistently distinctive barcodes (Hebert et al. 2004). A presumed generalist parasitoid wasp of the skipper, the braconid *Apanteles leucostigma* was later found to comprise 39 distinct species, and a group of tachinid fly parasitoids increased from 16 to 72 species. The general trend from such investigations has been that a presumed widespread generalist species is revealed to be a complex of several to many far more specialised species, each with far more restricted habits, including host or host plant range, than initially suspected. Many such cases may exist, and one problem they raise is that historical information on a 'broad generalist species' may be misleading and obscure the specialisations of then unrecognised constituent species.

Some insect orders (Appendix) are far more diverse than others: in Australia, they range from containing a single species (Zoraptera, known only from Christmas Island rather than mainland Australia) to many tens of thousands. The largest insect orders all have a complete metamorphosis (whereby larval and adult stages are essentially very different animals with different requirements: the herbivorous chewing caterpillar differs much from the corresponding winged nectar-feeding butterfly or moth into which it develops through an intermediate pupa stage, and a maggot and the adult fly into which it develops have very different appearance and capabilities). The more primitive insects have an incomplete metamorphosis (in which the larval stages resemble miniature adults, have similar feeding habits, and occupy the same habitats; wings develop externally, enlarging at successive moults). They are dominated by the true bugs (Hemiptera, including insects such as aphids, scale insects, cicadas, and plant bugs). All these large orders are predominantly terrestrial, but they also include taxa that have invaded aquatic ecosystems, so demanding independent specialisations for an aquatic larval way of life and for the adults usually being terrestrial. Some insect orders are strongly associated with freshwater environments; Odonata (dragonflies, damselflies), Ephemeroptera (mayflies), Plecoptera (stoneflies), and the more advanced Trichoptera (caddisflies) are the most characteristic of these and depend wholly on waterbodies for their larval life, as abundant components of freshwater communities.

Traditionally, beetles (Coleoptera) have been regarded as the most diverse order of insects, that with the greatest number of species—as reflected in an often-quoted response by the biologist J.B.S. Haldane who, when asked what one could conclude about the nature of God from a study of his creations, reportedly responded 'An inordinate fondness for beetles'. However, this predominance is open to question as more penetrating analyses of the diversity of small flies (Diptera) and parasitoid wasps (Hymenoptera) are made, enormous numbers of hitherto cryptic species are revealed and their documented richness increased markedly. It is still uncertain which of this trio is truly the most diverse. Essentially, all are 'megadiverse', and commentators have suggested that the long-held belief in the predominance of

Coleoptera has flowed from them being 'collectable', with beetle-collectors historically rivalling lepidopterists in their enthusiasms; beetles, being hard-bodied and relatively easy to preserve, have for long been popular cabinet inclusions. Tiny flies and wasps have never garnered equivalent collector interests. Lepidoptera occupy the fourth place in this numerical hierarchy. Together, these four orders are thought to comprise more than 80% of all insect species.

Some wholly or largely aquatic orders are significant components of freshwater communities. They, also, may have very restricted environmental needs and tolerances, and respond rapidly to changes. Ecologists have employed some characteristic aquatic insect orders in an index to monitor water quality: the 'EPT index' incorporates the richness and/or abundance of Ephemeroptera (mayflies), Plecoptera (stoneflies), and Trichoptera (caddisflies) as their larvae are amenable to standard comparative sampling. Dragonflies and damselflies (Odonata) are also included in some surveys (so that the index becomes 'EPTO')—but all these orders are sufficiently diverse for meaningful measures to be taken, based on clear species-level recognition of their aquatic larval stages and awareness that both local richness and assemblage composition are influenced by environmental conditions and change.

In contrast to the species-rich orders noted above, some others contain very few species, and some are of interest to only a few specialists. Many of these groups play little part in conservation planning (Appendix), despite their evolutionary and other significance. Understandably, most conservation focus devolves on the more diverse groups that furnish greater meaningful ecological information and perspective and with the reality that 'large numbers of species impress'. Insect conservation embraces all terrestrial and aquatic environments in which insects are found, but the strong taxonomic bias to larger and better-known orders reflects both traditional interests and scientific capability.

Descriptions of insect faunas in the northern temperate regions and (to a lesser extent) Australia and New Zealand benefit from resident entomological interest and expertise, and a social environment in which such activities are possible, so that continuing documentation and attention is possible. A combination of very high insect diversity with countries 'rich' enough to be able to heed their fate is indeed rare. The fates of many insect faunas are understandably neglected over much of the world as basic human needs prevail and food supply and health issues dominate individual and national concerns. Knowledge of the southern temperate insect faunas lags considerably behind that of the north, with the latter having far smaller and better-described faunas (Stewart and New 2007).

The wealth of insect life in the tropics still largely defies evaluation: the formidable problems of changing this in the face of severe threats to forests and other environments led Lewis and Basset (2007) to comment that, reflecting their intangibility, '… in practice, tropical insects will rarely, if ever, be the targets of conservation action in their own right'. Tropical forests harbour more insect life than any other ecosystems—but inventories to assess those insects are difficult to undertake, not least because the tall forest canopies so rich with insects are difficult to access, and much earlier work necessarily included only those insects found at near-ground levels. In addition, the numerous tree species present—sometimes well over 100 species/hectare—provide complex mosaics of resources and conditions fostering a great variety of insect life and rendering almost every forest patch unique, together with its resident insect community. One of Erwin's (1988) surveys of beetles collected over three seasons in two 'similar' forest plots only 50 m apart in Tambopata (Peru) showed that those plots had only 8.7% of species (n = 126) in common. Calls for All Taxa Biodiversity Inventory (ATBI) of tropical areas, initially for a site in Costa Rica (Janzen 1988), still appear largely impracticable, despite considerable more recent advances in methods and data storage and retrieval. As Erwin (1988) noted, the basic process of an inventory involves (1) finding all the species; (2) classifying them and rendering them recognisable; (3) making a voucher collection for permanent reference; and (4) storing the data in a readily retrievable form. The process is inevitably complex and time consuming; despite advances in methodology, Novotny and Miller (2014) noted that on average, a newly discovered tropical insect species will not be described formally for 20 years, so that 'tropical entomology remains both blessed and burdened with the diversity of species it purports to study'.

Detailed knowledge of the richness and ecological variety of tropical insect life remains sparse, not least because those areas mostly lack cohorts of resident entomologists with opportunity to undertake relevant studies, and access to specialists able to provide first-hand taxonomic identifications and guidance. Initial investigations of the forest canopy insect fauna revolutionised ideas on tropical insect diversity, with the paramount stimulus being a short note by Erwin (1982) interpreting his samples obtained by 'insecticide fogging' (using a biodegradable insecticide) from a single neotropical tree species in Panama. The technique involves hoisting a pesticide mist sprayer on ropes into the canopy, triggering its activity there and catching the falling rain of insects in funnels or on trays at ground level. Erwin's data (Table 1.1) allowed, for the first time, testing of hypotheses on distribution of insects amongst tropical trees in relation to host plant specificity, and stimulated similar studies elsewhere in the tropics (Stork 2018). His initial suggestion reflecting considerable host plant specificity, was that there could be as many as 30 million neotropical canopy arthropod species alone. This unexpectedly high figure led to considerable debate by introducing a figure far higher than those previously

Table 1.1 Summary of the method used by Erwin (1982) in estimating 30 million species of tropical arthropods (as displayed by Stork 1988)

1 Number of beetle species collected by fogging 19 individual *Luehea seemannii* trees.	1200
2 Average host specificity for beetles of 13.5%, so number of beetle species host-specific on *L. seemannii*	163
3 Number of tropical tree species is about 50,000, so number of beetle species host specific to these	8,150,000
4 Beetles represent 40% of canopy arthropod species, so number of species of tropical canopy arthropods	20,000 000
5 Canopy is twice as rich as ground in species of arthropods, so the total number of tropical rainforest arthropods	30,000,000

considered (it 'began the usual controversy of right or wrong': Erwin 1988). Erwin later suggested a figure of 80 million insect species. That debate continues.

The hundreds to thousands of insect species in even small tropical forest areas, and the generally lower (but still significant) richness in tropical freshwater and other terrestrial biomes sum to considerable biomass and also imply some form of 'importance', but thwart completion of ATBIs in favour of selecting particular target taxa for investigation (p. 102). Almost all insect surveys in the tropics yield large numbers of undescribed or previously unknown species—and many of those that can be named may be revealed to be members of 'species complexes' that differ in detail but look superficially alike. As Stork (2018) noted, many entomologists readily accept a global total in the range of 5–10 million species—but he also estimated that only around 20% of those species (using a total of 5.5 million species) had been named, so that high levels of uncertainty persist. From Novotny and Miller (2014), 'Our inability to enumerate tropical species has made estimates of their diversity popular', with a long history of extrapolating from plant diversity and estimated numbers of insects regarded as host-specific. However, the local spectrum of plant species may influence interpretation of specificity—for example only a minority of insect herbivore species may feed on a single plant species when alternative congeneric hosts are available (Novotny and Basset 2005).

Intrinsic to assessing insect diversity is that most insects are not ubiquitous but are restricted to particular environments and geographical regions. Thus, Erwin's calculations, above, assessed a level of 'host plant specificity', implying that species found on his single study tree species could not thrive on other hosts. Many insect herbivores are indeed restricted to feeding on single plant species—they are 'monophagous'—and relationships with those hosts can become intricate and obligatory, whether in forest, grassland, herbfield, or other vegetation categories. Others are less selective and can feed from many plant species, as being 'polyphagous'. At a higher trophic level, many insect predators and parasitoids are also very selective in their prey or hosts, and restricted to particular species and search arenas, whilst others are more generalised and widespread. These situations essentially represent an ecological gradient from 'specialists' to 'generalists', with extreme specialists depending wholly on a single food species and perhaps with their life history and seasonal development intricately linked with their food species. Responses to environmental changes in principle differ considerably; a generalist species may be able to compensate for loss of a particular resource simply by feeding on others, whilst the fate of a specialist is linked closely with a single food species without which it cannot survive. Unsurprisingly, many of the insects noted as of priority conservation concern are ecological specialists, threatened by declines of key resources and environments. Many are also very localised, reflecting distributions of critical plant species or other resources. Each specialist insect consumer species depends on the wellbeing and availability of its food, be it plant or animal, as well as other environmental conditions.

The functional roles of insects, whether based on specific interactions (such as pollinator-plant, p. 77) or more general themes such as being significant dietary components for many vertebrate and invertebrate consumers, and occupying key positions in complex food webs, are powerful reasons for sustaining them, and are discussed in Chap. 6. However, the roles of many species are unknown in any detail; despite a vast and increasing literature on insect interactions and roles within their communities, many taxa have not been studied, and the level of their ecological importance may never be clarified. They can be regarded as 'meek inheritors' whose wider functional and ecological importance remains unclear—but that cannot be discounted.

Most insect species, and many higher taxa (genera, families, and even some orders) are geographically restricted. Most Australian insects, for example occur nowhere else (so they are 'endemic'), and some are very characteristic elements of the fauna, rivalling the better-known birds and mammals as unique components of the country's biological heritage. Many insects are linked with the equally characteristic Australian flora and have coexisted (and in many cases coevolved) with particular plant species over long periods. The formidable botanical complexity of Australia's major radiations of eucalypts (Myrtaceae) and wattles (acacias: Fabaceae) is paralleled amongst the diverse insect assemblages restricted to them, but many other Australian endemic plants also harbour specialised locally endemic insect herbivores and their natural enemies.

Insect distributions are restricted also by climate. Most tropical insects do not extend into more temperate regimes, leading to complex distributional gradients along latitudes, elevations, and aridity. Diversity along any such ecological gradients can also vary continuously, and studies along gradients of anthropogenic change—such as urban gradients from town/city centres to nearby peripheral less disturbed or natural areas—often reveal changes in species richness and community composition, commonly with declines or disappearance of the more ecologically specialised insects in the most disturbed or most changed areas. Likewise, elevational gradients from lowlands to montane or alpine areas typically comprise a series of ecological 'zones', with each harbouring assemblages differing in detail from those above or below. Again, overall richness may decline upward under the harsher higher elevation regimes.

Different 'faunal elements' also intergrade. Australian native insects include (1) ancient Gondwanan forms, most characteristic of the cooler southern regions and with their closest relatives in other southern areas such as temperate South America, South Africa, and New Zealand; and (2) more northerly 'Oriental' taxa developing from waves of colonisations from (mainly) southeast Asia and New Guinea, so related to those faunas. These two groups show interchanges along the east coast, largely at higher levels on the Great Dividing Range or along the coastal fringe. Superimposed on this 'natural' fauna are the sometimes more cosmopolitan species arriving or introduced since European settlement. These include many of the more important pest species, exploiting resources also originating from afar; some have invaded more natural ecosystems and may compete with, and displace, native insect species. Such 'alien' or 'invasive' species can pose a severe threat to native insect species (p. 48).

Ecological and geographical distributions, both often very restricted, lead to many insects being characteristic of (and largely restricted to) particular ecosystems and areas. Thus, the grassland or forest areas of North America, Britain, Australia, and many other places, each support insect assemblages that may be broadly dominated by the same orders, but almost no species occur in more than one of these regions, and local evolution gives each such local fauna a very characteristic identity. However, external pressures on those grasslands or forests may be similar in different places, so that those insect faunas face parallel conservation needs—and management pioneered for any one region may provide useful lessons for application elsewhere. Grasslands, for example are partially managed by combinations of grazing, burning, and slashing/mowing in many parts of the world and the consequences of each—singly or in combination—may be relevant to other contexts and cases.

This distributional and ecological variety gives many Australian insects the status of being 'narrow range endemics' found only in very small areas in which their specialised resources are available. Some species are known only from single localities, or single small areas of suitable habitat. The precise locality of capture may remain unknown—labels on many older specimens in Museum depositories sometimes give only imprecise information such as 'Victoria' or, less widely, 'Grampians' or other regional indications of area. Further, many species have not been rediscovered since they were initially described, perhaps a century or more ago, and whether they still exist is an open question. Even intensive and persistent searches have not revealed some of these. The Black Mountain Reserve, close to the headquarters of the Australian National Insect Collection in Canberra, is one of the most intensively surveyed areas in Australia, with a large number of insect species recorded. Yet one tiny moth, *Lophocorona melanora* (Lophocoronidae), despite numerous searches, is still known from only a single individual captured there in 1949, when it was flying in the late afternoon. For this, and parallel cases spanning many insect orders, conservation needs cannot be assessed other than by urging caution over changes to the sites from where they were historically recorded, in the hope that they still exist. Such precautionary measures are needed widely in insect conservation, but are often difficult to render politically persuasive.

The picture of Australian insect life and its evolution continues to change with new discoveries, some clearly of global significance in elucidating insect evolution. The orthopteran known as the 'Cooloola Monster' formed the basis of a new family (as *Cooloola propator*, Cooloolidae) and initially appeared to its describer (Rentz 1980) so unusual that it could have been 'manufactured'—much as early scientists in Europe regarded the first specimens of platypus to reach them as artefacts! Some insects are regarded as 'living fossils', reflecting their primitive features. These include the Enigma moth (*Aenigmatinea glatzella*) from Kangaroo Island, South Australia, which also represented a new family (Aenigmatineidae) (Kristensen et al. 2015), as a very primitive species with larvae feeding on ancient *Callitris* conifers. Discoveries of both these examples, together with many others, created excitement amongst the entomologist community and demonstrated novel and unique components of the country's insect fauna. Both species remain poorly known, localised, and perhaps vulnerable to local environmental changes. *Aenigmatinea*, for example may be severely prone to loss through wildfire, with its *Callitris* food trees highly flammable.

In short, and again in marked contrast to Britain (despite caveats noted by Morris and Welch, 2023a, 2023b) inventories of insect species in Australia are rarely, if ever, complete, even for the best-known taxonomic groups. Numbers of species and individuals in an insect assemblage or population change continually, but many species retain some reasonably predictable level of incidence, from which changes may arouse concern. Some, for example are usually found in only small numbers and appear to be genuinely rare, whilst others are 'common'. More generally, ecologists and others have commented that 'rarity is very common' because many insect species are found in only very low numbers. Tropical arthropod samples not infrequently contain around 30% of singletons (below), and a similar level of species described from single specimens implies wide acceptance of the reality of rarity—but also of lack of ability to appraise intraspecific variation and the real boundaries between species (Lim et al. 2012).

The relative abundance of insect species, and their relevance to inventories and 'counting species', typically shows a pattern from samples in which few species are abundant and many more occur in low numbers. Often, 'singletons' (species represented by single individuals) and 'doubletons' (species with two individuals) outnumber all others and—although their presence is indeed confirmed—whether those species are genuinely rare, simply not captured, or are casual stragglers from other environments rather than resident is largely conjecture. Increasing the sample sizes typically supports this pattern—as more species are collected, more singletons accumulate and many of those previously present retain that status; whilst some may fortuitously transition to doubletons or other low numbers, their general low abundance is usually unchanged. Several extensive surveys of nocturnal moths across the world exemplify the incidence of singletons in samples: (1) Oregon forests, sampled over 5 years: 493 species included 61 singletons and 30 doubletons (Highland et al. 2013); (2) Malaysian tropical forests: 1426 species included 538 singletons and 366 species with two or three individuals (Barlow and Woiwod 1989): and (3) Australia's Wet Tropics Bioregion: 906 photographed morphospecies included 497 singletons (Tng et al. 2021).

Their significance is difficult to explain. For example, the species richness of herbivores on a plant species can be inflated by species moving from different neighbouring plants—and this may be particularly evident in rainforests where numerous plant species grow closely together (Novotny and Basset 2000). Novotny and Basset listed five categories of insect herbivores whereby rarity could be presumed: (1) transients, resting on a plant but not feeding on it; (2) specialists or generalists that are not sampled adequately by the method(s) used; (3) specialists with genuinely low population levels; (4) generalists feeding occasionally on the plant examined but for which overall population levels may be high when assessed across a wider host range; and (5) specialists feeding on the host examined but preferring and more abundant on other, related, hosts. The first two of these are essentially sampling artefacts, but are still difficult to assess, and the others have some real biological basis. In another overview, Miller et al. (2003) also recognised insufficient sampling and listed other possible reasons for moth 'rarity' as (1) larvae feeding on rare host plants, so local numbers are also limited; (2) species have narrow environmental tolerances; (3) influences of natural enemies; (4) strong competition with other species; and (5) edge distributions of the whole range of species.

'Rarity' is thus a somewhat subjective concept that embodies low abundance, ecological limitation, and narrow distribution but also reflects ease of detection and amenability to retrieval by methods used to sample or collect insects. It is a somewhat emotive term. Imprecision in defining it can hamper conservation through introducing ambiguity and, as Main (1982) commented, rarity can be used as a relative term simply meaning that some other species is 'more common'. To many people, rarity equates to desirability and increased value: 'rare butterflies' thus parallel 'rare stamps' in demanding increased prices from collectors. To many ecologists, rarity may reflect vulnerability and level of conservation interest through suggesting 'risk of extinction' as a key parameter of conservation status.

Australia's largest stick insects (Phasmatodea) have been described from their northern forests only in the last few years, and it may seem strange that such gigantic insects have escaped detection for so long, and that they are so poorly known. Females of the Gargantuan stick insect (*Ctenomorpha gargantua*) can exceed 58 cm in length, but it and others are believed to live mainly in the tall forest canopy and may come near the ground only after displacement by 'accidents' such as from severe weather. However, Brock and Hasenpusch (2009) suggested that, despite very few individuals having been captured (with one female of *C. gargantua* founding a captive-bred population at Museum Victoria), these insects may not be truly rare but simply elusive and, perhaps, occurring only at low densities.

Whatever the true state of 'rarity', the condition raises practical problems for survey and conservation assessment. 'Rare' insects are often the species of greatest conservation interest but: (1) may be very difficult to find in large or moderately large numbers; so (2) are costly to study, and data accrued may be very incomplete and not convincing for gaining future support; (3) it may not be feasible or desirable to capture specimens for detailed study, or for any form of experimental manipulation of the population; (4) it is sometimes uncertain whether the small populations known will persist long enough to provide the information needed for conservation; and (5) even when the biology is known sufficiently, chance environmental variations may thwart efforts at conservation (Main 1982; New 1996). In short, many rare insects are difficult to study, but a precautionary approach dictates that they should be considered genuinely scarce and vulnerable and that conservation measures may be desirable.

Chance discoveries or improved survey/detection methods may change impressions of rarity. The Golden sun-moth (*Synemon plana*, Castniidae) has become a flagship species for conservation of threatened native grasslands in southeast mainland Australia, and on which it was believed to be very rare with few populations recorded. Improved knowledge of the moth's biology and activity pattern (Gibson and New 2007) subsequently enabled detection of many more populations and led to a downgrading of its formal conservation status from 'Critically endangered' to 'Vulnerable'. Simply, many early surveys failed to detect the adult moths because they were undertaken under conditions and at times when the males were not active. The females fly little, and surveys must target males. These fly only during a few hours from late morning to early afternoon, and in conditions of low winds, sun, and no precipitation, within a restricted temperature span. Surveys at other times of day or in cool or inclement conditions cannot detect the moth, even though they may be present resting in grass tussocks or on the ground. As a further complication, each moth lives for only a few days, but with the population at any site emerging continuously over several months in late spring and early summer—so that even successful surveys can record only an unknown fraction of the resident population on any occasion. And it is still unclear whether the moth takes 1, 2, or even 3 years to complete development—so that the adults emerging in any season may represent only a cohort of a larger resident population.

Such behavioural and developmental vagaries of insect life are mostly not known, so that many 'spot surveys' may not be as reliable as anticipated or assumed! As other representative common biases: (1) many nocturnal moths fly predominantly only over characteristic parts of the night and (2) many insects are present and detectable over only a short emergence or flight season which may be governed in part by seasonal temperatures. Single samples may unknowingly detect low numbers near the beginning or end of that season, or large numbers at the time of peak activity.

Understanding insect behaviour is thus important in understanding incidence and distribution. Dispersal by flight, for example varies enormously across species, from long-distance migrants (in some cases over thousands of Km) to little local movement. The most intensively investigated migrant species is the Monarch butterfly (*Danaus plexippus*, Nymphalidae, in North America: the same species occurs in Australia where it is known as the Wanderer, and apparently arrived by 'island-hopping' across the Pacific). The butterflies overwinter in localised forest sites in California and Mexico, from where they fly northward to colonise much of North America over several generations through spring and early summer. In late summer, their descendants fly back to their 'traditional' overwintering areas (in which adults 'roost' on trees until spring) and in this case, protection of those sites from logging and wider despoliation and disturbance is essential for the butterfly to survive.

Counting insect species, however, they are delimited, spans considerations of time and space—and also assumes consistent (and accurate) recognition of the entities assessed. This very basic assumption is not always easy to validate, especially when comparisons across samples or sites are attempted, and when samples are analysed by different people. In many studies, voucher specimens of the taxa identified are not available for comparative study. Reflecting the dearth of entomological expertise (p. 147), the use of 'biodiversity technicians' has flourished, essentially as employing trained and supervised non-specialists to sort and recognise species in a given group and whose role is thus to accelerate sample processing and documentation. Errors can easily be introduced—through features such as the insect's sexual dimorphism, variations in colour or size, confusion over associations of early stages, and so on. Existing errors can also easily be undetected and perpetuated, and close monitoring by specialists is vital as 'quality control'. Different workers, even when expert in the same insect taxon, may come to different conclusions over separating species, and undescribed species provide further ambiguity. An almost universal logistic problem is that insect surveys yield enormous numbers of individuals, and it is rarely possible to examine all the taxa present. Estimates of total richness may be extrapolated from study of only selected taxa that, for reasons such as individual interest, ease of handling, and available expertise, are selected to in some way reflect or 'indicate' the wider diversity present.

The biology of almost all insect taxa found in a broad survey is likely to be largely or entirely unknown, or simply inferred to some slight extent. Host plant records may be available for some herbivores, for example but these are unlikely to be comprehensive and published records may, for example (1) perpetuate errors of plant identification; (2) fail to represent up-to-date taxonomy; (3) simply be mistaken; or (4) be misleading in that they report incidence of the insect on a plant without any evidence that it feeds there—it may, for example have fallen from different vegetation above or simply alighted there. Likewise, the full prey or host spectrum of insect predators or parasitoids is usually unknown and, as for herbivores, not associated clearly from samples. In general, the early stages of insects are not as well known or as easily identified as the corresponding adults. That many early stages of insects cannot be identified without rearing them to the adult stage restricts many sample analyses to the adult stages only. Except for some insect groups with aquatic larvae, notably Odonata (dragonflies and damselflies), Ephemeroptera (mayflies), Plecoptera (stoneflies), and Trichoptera (caddisflies), for which recognition of larvae to species level is often possible (reflecting attention to their values as indicators of environmental quality: p. 103), most insect inventories devolve solely on the adult stage. This is a major hindrance in assessing diversity: for example as Eisenhauer et al. (2019) noted 'the soil-dwelling larval stages of flying insects … have been woefully neglected … in conservation actions and policies'. Indeed, inventories of insect taxa have focused almost exclusively on those found at ground level and above. Although soil arthropod faunas are globally dominated by mites and springtails (Collembola), Rosenberg et al. (2023) estimated that ants and termites together constitute around 55% of total soil arthropod biomass (Fig. 1.2). Termites are predominant, representing about 40% of total biomass but with very uneven global distribution. Collectively, insects were estimated to comprise about two-thirds of total soil arthropod biomass. Estimating arboreal (or, more broadly, 'above-ground') biomass is far more tenuous: Rosenberg et al. found only two studies reporting arthropod biomass per area of forest floor, for example and few measured values exist for most other habitats. Nevertheless, from Rosenberg et al.'s estimates, terrestrial arthropods could number around 10^{19} individuals at any one time, with dry weight biomass of around 300 million tonnes—more than twice that of the total human biomass of around 120 million tonnes cited by Bar-On et al. (2018).

However, they are enumerated or otherwise quantified, insects comprise a large proportion of animal 'biodiversity', measured by species numbers or biomass, over much of the world. Clarifying the real extent of that diversity is a continuing challenge: it is unlikely that all insect species will be distinguished and/or formally

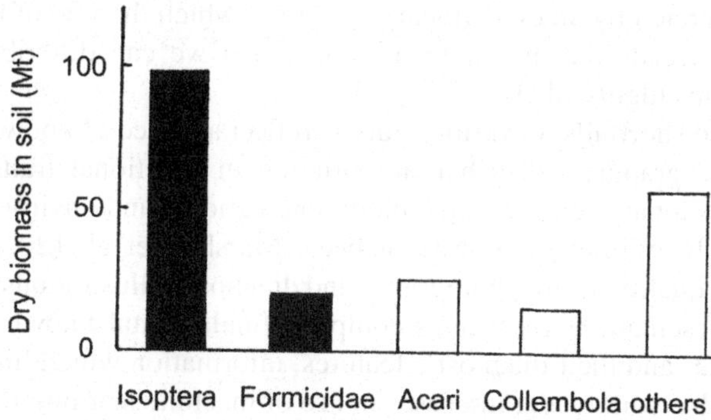

Fig. 1.2 The biomass distribution amongst the major groups contributing to the total biomass of soil arthropods (after Rosenberg et al. 2023). The two major groups of insects involved are shown as shaded columns (used under Creative Commons Attribution licence CC-BY-4.0)

Table 1.2 The seven impediments in invertebrate conservation discussed by Cardoso et al. (2011)

1.	The public dilemma. People throughout the world do not recognise invertebrates or their roles in ecosystems, so they are disregarded as objects in need of protection.
2.	The political dilemma. Many policymakers and stakeholders view invertebrates as species that, if needed, are indirectly protected by 'umbrella' vertebrate species, so that protection measures and funding are limited.
3.	The scientific dilemma. Taxonomy and classical ecology are underfunded, as increasingly regarded as founded on 'dated science'.
4.	The Linnean shortfall. The knowledge of insect species identity is poor, with many species yet to be described and catalogued.
5.	The Wallacean shortfall. Knowledge of distribution of species at all possible scales is inadequate.
6.	The Prestonian shortfall. Comparative species abundance data in space and time is usually scarce.
7.	The Hutchinsonian shortfall. Their diverse ways of life, functional roles and consequently the sensitivity to habitat change of most species are usually unknown.

described for centuries to come, and their importance and conservation needs must necessarily be clarified and communicated without that detail. This itself is a conceptual challenge to many biologists and managers who seek detail on which to base recommendations and activities. Insect conservation must consider, and incorporate, a number of impediments that illustrate lack of regard and a highly incomplete knowledge base. The seven such impediments recognised by Cardoso et al. (2011) comprise three 'dilemmas' and four 'shortfalls' applying across species and wider diversity (Table 1.2). These express both societal and scientific hindrances to advance, as foci for attention to improve current limitations of understanding. Other complexities of 'shortfalls' have been advanced, but all reflect the difficulties of trying to understand the gaps between actual knowledge and the desirable complete knowledge that would enable fuller understanding. Hortal et al. (2015) suggested that shortfalls in biological knowledge might be 'a direct consequence of the complexity generated by an evolutionary system in which the rate of production of novel entities exceeds the maximum rate at which we can describe them'. For insects, this seems highly likely.

The identified shortfalls, in various guises, reflect absence of knowledge on species identity, geographical distribution, variation in functional traits, population dynamics, evolutionary relationships, biotic interactions, and environmental tolerances—to which, writing on European bees, Marshall et al. (2024) added the absence of high quality in situ photographs and diagnostic illustrations. They argued that visual representations of species comprise fundamental knowledge of 'what species look like' and their diagnostic features, information which links with other shortfalls, but also with recent increases in use of machine learning tools for insect identification and citizen science recording sites such as iNaturalist (p. 57).

In short, whilst insects are almost endlessly diverse, insect taxonomists are not. Overriding many of the practical problems of interpretation indicated in this chapter is simply the paucity of taxonomic expertise available to identify and describe insect species and produce the reference materials—such as handbooks—needed by field biologists to enable them to interpret insect samples and diversity. Many existing keys and synopses are out of date, and need revision to incorporate the species discovered since they were written; it follows that in many cases their continued use may introduce errors of identification. Most species of insects—with exceptions such as many butterflies and distinctive species of other groups—will not be easily recognisable in the field, and close examination may be necessary to be sure of their identity. This situation raises a deep ethical dilemma, in that most surveys of insects involve sampling by lethal means, and sorting the accumulated catches for the taxa of interest, in many cases leaving a considerable amount of unwanted 'bycatch' that is never analysed and is often discarded for want of storage space or immediate interest. Such uncritical slaughter of insects is ethically difficult to condone as part of a conservation exercise, and does little to promote public sympathy for such activities. This scenario, however, is an integral facet of insect surveys, the source of much-needed voucher material for taxonomic and faunistic analyses. In some contexts, non-lethal alternatives may exist. Lovei and Ferrante (2024) have urged that these should be considered at every opportunity in order to lessen unwarranted mortality.

In contrast to the uncertainties of many insect surveys, conservationists working with vertebrates can usually recognise the species with which they work, as well as other taxa. Despite some problems with similar-looking taxa, surveys of these taxa usually have little ambiguity—discoveries of entirely new species are relatively uncommon, and well-illustrated identification handbooks are available for many parts of the world. In general, most bird observers and mammal specialists can recognise and identify all or most of the species they encounter. And many surveys for vertebrate diversity include rather few species. In contrast, insect surveys can yield hundreds of species, many of them small and very similar, and across a variety of orders for which identification of species is impracticable except by a specialist—who may still not be able to provide 'names'. Insect taxonomists can spend their entire working careers studying a single family (or even genus) of beetles, flies, or wasps, and not complete the task. Species recognition may depend on small morphological features visible only after careful preparation and microscopic examination. Insects have many equivalents to the ornithologist's 'little brown bird': the eponymous 'little black beetle' or 'tiny shiny wasp', for example may encompass representatives of many families with varying biologies and conservation needs, and which are not easily differentiated. These difficulties have driven conservation emphasis toward insect groups in which species are most easily recognised—well-illustrated handbooks for larger Lepidoptera, Odonata, and some beetle families (their development flowing from collector interests) remove much ambiguity of applying species names because many of the named species can be diagnosed on features of colour pattern or other conspicuous characteristics, but even then complexes of similar species need closer examination—either by capture and release or retention of voucher specimens for more detailed study. For many other orders, families and (sometimes) genera can often be recognised unambiguously but species recognition is laborious and uncertain, adding substantial difficulty and cost to interpreting field samples. Over much of the tropics and other under-explored regions, gaps in taxonomic understanding far exceed the current capability to address them.

A recent survey of insect taxonomists in Europe revealed that taxonomic capacity is itself threatened and was absent for some orders. Hochkirch et al. (2022) noted that decline of taxonomic expertise in parallel with insect declines erodes the research base for conservation and many other fields. Even for relatively well-known insect taxa in the European fauna, major gaps in knowledge persist, and some orders are sorely in need of taxonomic attention, which may best be remedied at the 'country level', by securing support for major museums and related institutions. This report, presented as a 'Red List of Insect Taxonomists', used categories equivalent to the threat categories used for species in the IUCN Red List (p. 112) with 'Eroded Capacity' (equivalent to 'Extinct'), 'Critically Low Capacity' ('Critically Endangered'), 'Inadequate Capacity' ('Endangered'), and 'Poor Capacity' ('Vulnerable'). For Europe, 'Adequate Capacity' (equivalent to 'Least Concern') was found for only 14 of the 29 insect orders, with 12 orders in the above three 'Threatened' categories. In many cases, and reflected elsewhere in the world, taxonomic capability was correlated strongly to a country's Gross Domestic Product. Soberingly, the Red List also reports that the insect taxonomist complement is aging, and relatively few young people are entering the profession.

Simply 'describing insect species' is often not seen as an attractive (marketable!) career option. Recent recruits to the declining entomologist workforce in even leading global museums often seek to investigate phylogenies and allied evolutionary relationships within insect groups by using sophisticated statistical and molecular methods (as transferable skills that can be used across different taxa), rather than delineating species in terms that transfer easily to the real world of field sample interpretations and in which species' identifications are the working currency for analysis. Indeed, opportunities for young people to become practicing insect taxonomists are themselves diminishing as training in basic 'classical' entomology disappears from many tertiary courses in biology (p. 144). Difficulties flow particularly to poorly known tropical faunas, most in countries with very few resident insect taxonomists and in which species identifications may depend on comparisons with historical material deposited in European museums in which staffing is eroding, and for which the limited staff time has to be reimbursed. Whilst newly developed methods of species enumeration and distinction—many based on machine learning or molecular approaches and continually advancing in scope and subtleties—are helping enormously to characterise species and faunas, simply being able to 'recognise insects' and identify them reliably is a core need for many workers in insect conservation.

An important manifestation of insect diversity and abundance is simply the 'amount' of insects—their biomass—as another axis of significance and change. Loss of biomass, as indicated by changes in the bulk of insects in trap catches over time (Hallmann et al. 2017, p. 23) reflects broad-scale concerns that transcend those from identified taxa alone to suggest far broader losses from local faunas. 'Wet biomass' of trapped insects has been used increasingly in insect studies, and allows the samples to be preserved for later more detailed examination. Kothe et al. (2023) also used insect metabarcoding to produce qualitative species lists, and direct examination of specimens by experts to evaluate changes in insect diversity in conservation areas.

References

Barlow HS, Woiwod IP (1989) Moth diversity of a tropical forest in peninsular Malaysia. J Trop Ecol 5:37–50

Bar-On YM, Phillips R, Milo R (2018) The biomass distribution on earth. Proc Nat Acad Sci 115(25):6506–6511. https://doi.org/10.1073/pnas.1711842115

Braby MF, Eastwood R, Murray N (2012) The subspecies concept in butterflies: has its application in taxonomy and conservation biology outlived its usefulness? Biol J Linn Soc 106:699–716. https://doi.org/10.1111/j.1095-8312.2012.01909.x

Brock PD, Hasenpusch JW (2009) Complete field guide to stick and leaf insects of Australia. CSIRO Publishing, Melbourne

Cardoso P, Erwin TL, Borges PAV, New TR (2011) The seven impediments in invertebrate conservation and how to overcome them. Biol Conserv 144:2647–2655. https://doi.org/10.1016/j.biocon.2011.07.024

Duffus NE, Morimoto J (2022) Current conservation policies in the UK and Ireland overlook endangered insects and are taxonomically biased towards Lepidoptera. Biol Conserv 266:109464. https://doi.org/10.1016/j.biocon.2022.109464

Eisenhauer N, Bonn A, Guerra CA (2019) Recognizing the quiet extinction of invertebrates. Nature Comm 10:50. https://doi.org/10.1038/s41467-018-07916-1

Erwin TL (1982) Tropical forests: their richness in Coleoptera and other arthropod species. Coleopt Bull 36:74–75

Erwin TL (1988) The tropical forest canopy. In: Wilson EO, Peters FM (eds) Biodiversity. National Academies Press, Washington DC, pp 123–129

Gibson L, New TR (2007) Problems in studying populations of the golden sun-moth, *Synemon plana* (Lepidoptera: Castniidae), in south eastern Australia. J Insect Conserv 11:309–313

Hallmann CA, Sorg M, Jongejans E, Siepel H, Hofland N (et al.) (2017) More than 75 percent decline over 27 years in total flying insect biomass in protected areas. PLoS One 12; e0185809.https://doi.org/10.1371/journal.pone.0185809

Hebert PDN, Penton EH, Burns JM, Janzen DH, Hallwachs W (2004) Ten species in one: DNA barcoding reveals cryptic species in the neotropical skipper butterfly *Astraptes fulgerator*. Proc Nat Acad Sci 101(410):14812–14817. https://doi.org/10.1073/pnas.04061666101

Highland SA, Miller JC, Jones JA (2013) Determinants of moth diversity and community in a temperate mountain landscape: vegetation, topography and seasonality. Ecosphere 4(10):129. https://doi.org/10.1890/ES12-00384.21

Hochkirch A, Casino A, Penev L, Allen D, Tilley L et al (2022) European red list of insect taxonomists Publications Office of the European Union, 2022. https://data.euripa.eu/doi/10.2779/364246

Hortal J, de Bello F, Diniz-Filho JEF, Lewinsohn TM, Lobo JM, Ladle RJ (2015) Seven shortfalls that beset large-scale knowledge of biodiversity. Annu Rev Ecol Evol Syst 46:523–549. https://doi.org/10.1146/annurev-ecolsys-112414-054400

Janzen DH (1988) Tropical dry forests. The most endangered major tropical ecosystem. In: Wilson EO, Peters FM (eds) Biodiversity. National Academies Press, Washington DC, pp 130–137

Kothe S, Bakanov N, Bruhl CA, Eichler L, Fickel T (et al.) (2023) Recommendations for effective insect conservation in nature protected areas based on a transdisciplinary project in Germany. Environ Sci Europe 35: 102. https://doi.org/10.1186/s12302-023-008135

Kristensen NP, Hilton DFJ, Kallies A, Milla L, Rota J (et al.) (2015) A new extant family of primitive moths from Kangaroo Island, Australia, and its significance for understanding early Lepidoptera evolution. Syst Entomol 40: 5–16. https://doi.org/10.1111/syen.12115

Lewis OT, Basset Y (2007) Insect conservation in tropical forests. In: Stewart AJA, New TR, Lewis OT (eds) Insect conservation biology. CAB International, Wallingford, pp 34–56

Lim GS, Balke M, Meier R (2012) Determining species boundaries in a world full of rarity: singletons, species delimitation methods. Syst Biol 61:165–169. https://doi.org/10.1093/sysbio/sys030

Lovei GL, Ferrante M (2024) The use and prospects of nonlethal methods in entomology. Annu Rev Entomol 69:183–198. https://doi.org/10.1146/annurev-ento-120220-024402

Main AR (1982) Rare species; precious or dross? In: Groves RH, Ride WDL (eds) Species at risk: research in Australia. Australian Academy of Science, Canberra, pp 163–174

Marshall L, Leclercq N, Carvalheiro LG, Dathe HH, Jacobi B (et al.) (2024) Understanding and addressing shortfalls in European wild bee data. Biol Conserv 290: 110455. https://doi.org/10.1016/j.biocon.2024.110455

Miller JC, Hammond PC, Ross DNR (2003) Distribution and functional roles of rare and uncommon moths (Lepidoptera: Noctuidae: Plusiinae) across a coniferous forest landscape. Ann Entomol Soc Am 96:847–855

Morris R, Welch M (2023a) Institutional vertebratism is alive (and kicking invertebrates). Antenna 47:23–27

Morris RKA, Welch MD (2023b) Is invertebrate conservation in Great Britain best achieved by policies that increase species protection? J Insect Conserv 27:527–531. https://doi.org/10.1007/s10841-023-00485-9

New TR (1984) Insect conservation, an Australian perspective. W. Junk, The Hague

New TR (1996) Evaluating the status of butterflies for conservation. Decline and conservation of butterflies in Japan III. Proc Int Symp Butterfly Conservation, Osaka, Japan 1994. Yadoriga, special issue: 4–21

Novotny V, Basset Y (2000) Rare species in communities of tropical insect herbivores: pondering the mystery of singletons. Oikos 89:564–572

Novotny V, Basset Y (2005) Host specificity of insect herbivores in tropical forests. Proc R Soc B 272:1083–1090. https://doi.org/10.1098/rspb.2004.3023

Novotny V, Miller SE (2014) Mapping and understanding the diversity of insects in the tropics: past achievements and future directions. Austral Entomol 53:259–267. https://doi.org/10.1111/aen.12111

Rentz DCF (1980) A new family of ensiferous Orthoptera from the coastal sands of Southeast Queensland, Australia. Mem Qld Mus 20:49–64

Rosenberg Y, Bar-On YM, Fromm A, Ostikar M, Shoshany A (et al.) (2023) The global biomass and number of terrestrial arthropods. Sci Adv 9, eabq4049

Sterling MJ, Plant CW, Price BW, Cuber P, Hall A (et al.) (2023) *Tachystola mulliganae* sp. nov. (Lep.: Oecophoridae), a Western Australian species adventive in West London. Entomol Rec J Var 135: 225–243

Stewart AJA, New TR (2007) Insect conservation in temperate biomes: issues, progress and prospects. In: Stewart AJA, New TR, Lewis OT (eds) Insect conservation biology. CABI, Wallingford, pp 1–33

Stork NE (1988) Insect diversity - facts, fiction and speculation. Biol J Linn Soc 35:321–337

Stork NE (2018) How many species of insects and other terrestrial arthropods are there on earth? Annu Rev Entomol 63:31–45. https://doi.org/10.1146/annurev-ento-020117-043348

Tng DYP, Apgaua DMG, Fisher NJ, Fazzio VW III (2021) Moth species richness in an upland tropical rainforest: a citizen scientist assisted study. https://doi.org/10.1101/2021.11.07.467659 (preprint posted 8 Nov 2021)

Yeates DK, Harvey MS, Austin AD (2003) New estimates for terrestrial arthropod species-richness in Australia. Rec S A Mus, Monogr series 7:231–241

Chapter 2
Insect Declines

2.1 Introduction

Losses—both severe declines and extinctions—of insects are not a new phenomenon, but until quite recently concerns have focused largely on the fate of individual species rather than on wider diversity and consequences of its loss. Single-species focus is not always feasible, but has been the major avenue along which insect conservation has been developed, mainly in the northern temperate regions with relatively small insect faunas and a high insect species: entomologist ratio. There, intensive study of individual species and their ecological needs can furnish well-informed conservation measures tailored to those needs. In contrast, the massive insect diversity of insects in tropical forests and the few entomologists able to appraise their fate, largely precludes this approach and dictates that the main focus is on changes within complex assemblages (Lewis and Basset 2007). The conservation of insect diversity reflects the central importance of the habitats in which that richness occurs.

Classic historical cases of butterfly losses were the major stimulus for developing conservation interest, with some programmes spanning decades or more of concern. The loss of the Large copper (*Lycaena dispar dispar*) from the fens of eastern England as they were lost to drainage in the mid-nineteenth century, and of the Large blue (*Maculinea arion*) more recently in Britain, both led later to successful reintroduction programmes (p. 124) to establish closely related subspecies from mainland Europe in Britain, and are among the all-time classic 'success stories' of insect conservation. A third lycaenid, the Xerces blue (*Glaucopsyche xerces*), finally became extinct in the United States in 1943, after fears of loss since the 1880s (Pyle 2012). Ecologically specialised Lycaenidae have continued to be significant 'flagships' for insect conservation in many parts of the world, as susceptible to many aspects of habitat loss and intricate associations with particular larval food plants and, for many, specific mutualist ants. In Australia, the Eltham copper (*Paralucia*

T. R. New, *Insect Conservation in Australia: Why and How*,
https://doi.org/10.1007/978-3-031-66631-5_2

pyrodiscus lucida) had a parallel catalysing role (New 2011) and was among the first focused conservation cases in which a local community became involved. The intricate tripartite associations of many lycaenids (that is of butterfly—specific food plant—specific host or mutualist ant) have for long fascinated biologists but also create vulnerability if any partner is affected by environmental change. Adding to this complexity, some Lycaenidae are known to support specific wasp parasitoids that may be more threatened than their hosts—and even treated as an undesirable threat to the butterfly.

Not all losses have been of such specialised, and thereby vulnerable, species. The North American Rocky Mountain locust (*Melanoplus spretus*) was abundant and widespread throughout the western United States and Canada in the nineteenth century—one swarm in 1875 was stated to extend over 510,000 Km^2. The locust was last seen alive in 1902, and its rapid decline and demise have been attributed to ploughing of the oviposition areas as agriculture expanded following the influx of gold miners in the region (Lockwood 2004). These examples, and numerous others, exemplify the sensitivity of many insects to environmental conditions and the ease with which suitable conditions can be eroded. They are simply a few conspicuous representatives of what is almost certainly far wider and unheralded loss of species wherever insects live. Especially in parts of the tropics undergoing rapid habitat changes such as forest clearing and wider agricultural conversion, many insects and other invertebrates may be lost before they are even known to humanity—let alone named and their biology understood. This unknown flow of extinctions creates considerable uncertainty over the extent of 'real' losses: thus Dunn (2005) could document only 70 modern insect species extinctions 'although thousands are estimated to have occurred', with losses most evident among ecologically specialised and locally endemic species and their close interdependent associates such as mutualists. The suggestion of 'thousands' of species likely to have been lost—impossible to validate or confirm but equally that cannot be refuted—drives major concerns for the impacts of human activities on insect faunas.

Insect species declines are the most visible part of the much wider overall losses of insect richness, abundance, and biomass, long heralded but relatively slow to garner attention. Renewed concerns over apparently severe declines of insects have come to the fore over the last decade or so, with fears expressed through many claims that insects are indeed suffering greatly as the world changes, with numerous extinctions and more local losses (extirpations) ranging from single species to faunal erosion spanning many taxa. It is sometimes difficult to separate hyperbole—however, well-founded in genuine concern—from reality, and debates continue over the extent of recent and likely future losses. Emotive (and sensationalistic) terms such as 'Insectageddon' (Leather 2019) and 'The Insect Apocalypse' (Goulson 2019) express the fears of massive (catastrophic) losses, and appeal to public and political constituencies alike—and, whilst some levels of loss are undoubted, the massive tragedies invoked by the above (and similar) terms are more difficult to validate, with claims and counterclaims expressed repeatedly, and 'proof' elusive. In reality, a 'decline' generally refers to a temporal pattern showing a downward trend of insect diversity or abundance over repeated (usually annual) measures at the same site, or such negative trends across locations (Bluthgen et al. 2022). More commonly, they are inferred from change across a pair of samples taken at intervals and perhaps under different conditions and methods rather than from standardised periodical measurements.

However, there is wide consensus that (1) insect declines are genuine and widespread across taxa, habitats, and regions; (2) they are difficult to quantify convincingly; (3) they are likely to have serious ecological and practical consequences; and (4) we need to know more about them and how to slow them and prevent them from proliferating. As Wagner (2020) put it 'with some important exceptions, the phenomenon of insect decline is both poorly documented and poorly understood'. Few entomologists doubt that insects are in various ways declining throughout the world, and that those declines—including those of species once regarded as 'common'—have serious ecological consequences, for example in eroding the food sources of birds and other consumers, and reducing fundamental processes such as pollination and decomposition. Declines have been inferred and documented most amongst the better-known faunas of the northern hemisphere but Australia and other southern realms are by no means immune.

Much of the recent alarm and concern flowed from a review by Sanchez-Bayo and Wyckhuys (2019), who assembled many of the important studies of the drivers of insect declines, but whose methodology has been criticised (Thomas et al. 2019). Nevertheless, studies such as the above and that by Hallmann et al. (2017) showing loss of 75% of insect biomass in Malaise trap catches in Germany over the 26 years from 1989 to 2014 are sobering, and the Hallmann conclusion is based on one of very few available European data sets of differences over such extended periods. Most of the examples noted by Sanchez-Bayo and Wyckhuys are from western Europe or North America, where most information has been accumulated, but with the additional cases considered in their later account (Sanchez-Bayo and Wyckhuys 2021) extending the inferences to include 100 long-term studies across many parts of the world. For most other places and most insect groups any such trends cannot be based on sound data. Occasionally they may be inferred indirectly—for example, Goulson (2019) noted declines in insectivorous birds in Europe and North America relating to putative declines in insect food. In contrast to such developed countries, in the Luquillo Experimental Forest of Puerto Rico, comparisons of sweep-net and sticky trap samples showed declines in arthropod biomass by 4–8 times (sweep-net) and 30–60 times (sticky traps). Lister and Garcia (2018) attributed these losses to effects of climate warming (p. 59). This claim caused wide interest, but also claims that the interpretation may be flawed (Showalter et al. 2021). A study from China (Zhou et al. 2023, p. 28) confirms that declines are geographically widespread.

However, much of the wider publicity arose from three now widely cited accounts noted above (Hallmann et al. 2017; Lister and Garcia 2018; Sanchez-Bayo and Wyckhuys 2019) but was initiated from widely distributed media releases from the host institutions, in turn leading to a flurry of media articles and extrapolated claims of the levels of loss—even to the extent of claiming that all insects could be lost over the next century. Those claims were repeated uncritically in a number of subsequent accounts. Calls for 'evidence-based' rather than more emotional inferences of declines continue, in the expectation of leading to a more realistic picture. Despite the positive values of drawing widespread attention to insect losses, the spread of 'misinformation' (in the traditional phrase 'crying wolf') through hyperbole or inaccuracy, however, well-intentioned, can easily backfire and alienate the badly needed societal sympathy on which insect conservation increasingly depends. Didham et al. (2020) noted that the intensive media interest might indeed be a 'double-edged sword' if the details of purportedly alarmist reports do not meet critical scrutiny. Seven key challenges were nominated as important in attempting to establish the reality of declines. Nevertheless, 'spreading the message' of concern over insect declines, and that uncertainties over their rates and extents do not mask that they are indeed occurring, is a vital plank in the conservation platform. Thought-provoking essays written for a wide readership, notably those by Goulson (2021) and Milman (2022), emphasise the need for increasing public awareness of insect roles and vulnerabilities.

This chapter includes a brief discussion of the debates over the extent and reality of insect declines as a prelude to evaluating possible causes and remedies. But, again from Wagner (2020), 'much uncertainty remains' in appraising the rates, extents, and consequences of insect loss, with precipitous declines of some previously abundant species perhaps having the most serious ecological impacts.

2.2 Assessing the Evidence for Insect Declines

Loss or decline of individual species is the historical foundation of insect conservation interest, but trends for wider insect biodiversity are more complex to investigate—simply because the requisite data do not exist. Extents of past losses are difficult to confirm, but concerns over some well-known species (especially of butterflies) extend back to the nineteenth century—and it is from them that 'insect conservation' interest had its birth. Records are uneven and often incomplete, but accumulated historical information allows detection of trends in incidence and abundance. For butterflies of the Netherlands, records from 1890 to 2017 implied an estimated decline of 84%, with downward trends for 42 of the 71 species evaluated (van Strien et al. 2019). Declines occurred in all three main butterfly habitats, grassland, woodland, and heathland.

Clearly, extinctions have always occurred as environments change, and their past levels are unknown. In principle, a species has three possible options when faced with its environment becoming unsuitable—it may be able to adapt to the changes and remain, it may be able to disperse to other still suitable sites, or it may become extinct. Ecologically specialised species are most likely to be influenced by environmental changes because they are unable to adapt to changes. Their replacement by increased numbers of more generalised species with wider resources available leads to increasing community homogenisation (Gossner et al. 2023) in which overall abundance may not be changed but species composition is markedly different—emphasising the need for species-level evaluations to determine the real impact and change that has occurred. Attempts to predict which insect species might be most susceptible have invoked consideration of a wide array of 'traits'—for example, low dispersal capability may be associated with vulnerability, whilst good dispersers may escape. However, the variety and idiosyncrasies of insect ecology make any such appraisal difficult, with population trends varying with taxon and local context. Likewise, several studies have implied that losses of terrestrial insects may be greater than declines of aquatic insects, a circumstance that may indeed be genuine but is also accompanied by relative lack of long-term monitoring in many places. Indeed, in some cases freshwater insect abundances may have increased while terrestrial insects have declined (van Klink et al. 2020); changes in richness and in abundance must be distinguished carefully.

Recent concerns have emphasised that the extent of insect losses is far greater than previously supposed, and that relatively recent anthropogenic pressures are largely responsible for this accelerated rate of loss. Few entomologists doubt that substantial losses of insects have occurred, that rates of loss are accelerating, and that many ecosystems they frequent will themselves be changed irreparably as a result, both from lowered diversity and changed ecological interactions. Insect declines have major significance for humanity and the world's biota.

Major fears emphasise, in part, that (1) the first of the above options (adaptation) has diminished or vanished for many insects because of large-scale environmental changes, (2) the second option (movement) is also thwarted through loss and increased isolation of many suitable habitat patches, and that (3) the third option (extinction or extirpation) then becomes increasingly likely. Extinction occurs when the last individual(s) of a species die, but some scientists recognise other forms of the term—'functional extinction' occurs when the species becomes too scarce to perform its former ecological role, and 'commercial extinction' occurs when the species is so scarce that it can no longer be harvested economically. 'Extirpation' refers to local extinction, such as populations of species that still occur elsewhere. 'Declines' reflect lessening of abundance and/or distribution. Many older naturalists recalling their younger days have recollections of insects being more 'common' than they now appear to be. In Australia, declines of large spectacular insects such as the Emperor gum moths (*Opodiphthera eucalypti, O. helena*, Saturniidae), Christmas beetles (*Anoplognathus* spp., Scarabaeidae), and some cicadas in urban areas arouse comment, for example but such reminiscences remain subjective—although repeated frequently by concerned naturalists. Proving these or more severe declines convincingly requires objective and long-term data that is only rarely available—so that an element of subjectivity can weaken advocacy. However, trends such as the reduced numbers of insects attracted to light or squished on vehicle windscreens or in radiator grills have been reported widely across the northern hemisphere and (although the latter have been attributed in part to better vehicle design) are difficult to dismiss. Janzen and Hallwachs' (2019) essay, drawing on decades of personal observation in Costa Rica and elsewhere, leaves little room for doubt that these trends are genuine and represent true declines in both abundance and species richness across a wide array of ecosystems. And, soberingly, they could not see any group/s of insects that were immune to those changes, so that the whole array of interactions to which insects contribute may also be affected. Nevertheless, it is unreasonable to expect any uniformity or consistency of pattern or trend across the numerous different insect groups—or even across species in the same group—or in different places. Inter-specific differences in resource needs, phenology, tolerances to disturbance, dispersal prowess, and other attributes ensure varied responses to a changed environment—with 'winners', 'losers', and others largely unaffected. Local variations in environmental conditions and threats also affect patterns of decline—a winner at one site may become a loser at another. All of the studies reviewed by Sanchez-Bayo and Wyckhuys (2019) implied this variety—as Saunders et al. (2020) pointed out (1) some of those studies found no declines in focal taxa and (2) in many, more taxa either increased or remained constant than declined. Dicks et al. (2024) stated that the 'evidence base' for decline in insect biodiversity had three major weaknesses, as taxonomic bias, spatial bias, and short time-series. The last of these has led to consideration of alternative approaches, such as investigating whether spatial comparisons might compensate for or augment paucity of temporal comparisons (Bluthgen et al. 2022). Many published studies are on Lepidoptera (major taxonomic bias) in parts of Europe and North America (spatial bias, but most studies represent ecosystems that have been modified extensively), with a widespread lack of long-term historical data (short time-series, and often not assessing the natural fluctuations in insect populations). The last is rarely considered but if a species' monitoring sequence starts at a time of naturally high abundance, subsequent declines might be overestimated (Dicks et al. 2024). Many insect populations can differ in size by up to several orders of magnitude across sequences of generations, and generation times can be short (even a few weeks, with some aphids and others far shorter) and the pattern of that variation is unknown for most taxa. Population size reflects both birth rate and mortality rate, both influenced by multitudinous and variable factors. Even standard intergenerational monitoring of univoltine species (such as annual surveys used for many univoltine butterflies) may inadvertently mask such trends. Three key areas nominated by Didham et al. (2020) for focused future attention are baselines (is the starting point representative?), trends (nature and extent), and population-level understanding.

The extensive and long-term monitoring information on British Lepidoptera, with some less extensive schemes elsewhere in western Europe, furnishes the strongest available data set against which to measure change in detail. Extensive networks of light traps, under the initial aegis of the Rothamsted Insect Survey, allow objective interpretation of species losses and gains amongst nocturnal moths, and of changes in their abundance and distribution. In contrast to many purported monitoring studies, methods of sampling have remained consistent over the period: in many other studies, it is unknown to what extent sampling methods changed in form or intensity, and their effects in interpreting the results. Some British moth species have indeed declined severely but others have increased. However, many moths previously regarded there as 'common' have declined markedly (Conrad et al. 2006; Fox 2013; Fox et al. 2014, 2021) with some of the main inferences summarised in Table 2.1 to illustrate the emerging picture—and also the difficulties of ascribing clear causes to such changes even in this best-documented fauna, in which the

Table 2.1 Some of the key features of documenting the changing picture of larger British moths, based on long-term light-trap catches: information extracted from the references cited

Conrad et al. 2006	Data from the 337 larger moth species represented by more than 500 individuals over the 35 year (1968–2002) trapping period; two-thirds of species declined over this period, severe population declines (>30%/10 years) found for 71 species, many of them 'common and widespread', would lead to a formal assessment as 'threatened'.
Fox et al. 2014	Used a larger data set to examine distribution and extent of changes: a total of 673 macro-moth species over the period 1970–2010. Overall, 260 species declined significantly and 160 increased significantly. The complex interplay between species' attributes and combinations of drivers of change (land use, climate) emphasised the needs to provide conservation for existing populations and to enable them to shift ranges in response to climate influences.
Fox et al. 2021	The most comprehensive appraisal so far, summarising knowledge of ca 900 larger moth species, over 1968–2017; total abundance of moths caught decreased by a third over this period, with losses greater in the southern half of Britain (39% decrease) than in the northern half (22%). For 427 species used to assess abundance trends 175 (41%) had decreased, 42 species (10%) had increased and the remainder (210 species, 49%) showed no statistically significant change. Distribution trends were: 165 species (32%) decreased, 37 species (37%) increased, and 159 species (31%) had non-significant trends. The number of moth extinctions (including micromoths) in Britain since 1900 was given as 51 species; 137 species arrived in Britain over that period, from natural colonisation from mainland Europe or inadvertent introductions through the plant trade.

moths still vary greatly in their biologies and species' needs. Fox (2013) commented that 'Direct evidence for causes of moth declines is extremely limited', but that the available studies implicated habitat degradation and climate change as likely to be the major drivers. Such long-term data series can counter the almost universal short-term changes due to normal fluctuations in population size, and avoid misinterpreting such shorter transient trends.

The declines of previously abundant species are of particular concern, because these may have considerable functional importance; in Wagner's (2020) words, they may 'anchor trophic interactions and shoulder many of the essential ecosystem services' in the communities they occupy. Simplistically, rare species may not have such influence, so that their loss may be functionally less significant—although typically of greater concern in conservation. But, whatever their functional significance, 'rare and uncommon species with special or restricted habitat requirements provide an important contribution to the biodiversity within a local landscape' (Miller et al. 2003, writing on moths). One recent concern in Australia has been the decline of the previously abundant iconic Bogong moth (*Agrotis infusa*, Noctuidae), leading to its designation as Endangered by the World Conservation Union in 2021 and attracting wide media publicity. The moth undergoes spectacular migrations from its lowland breeding sites on the southeast mainland to aestivate in caves in the restricted alpine region, and over the last decade or so, numbers of migrating moths have fallen from an estimated 4.4 billion individuals before 2017 to only a few thousand in later years.

However, few such parallel long-term data to the British Lepidoptera occur elsewhere—but few places are immune to insect declines and their impacts, with rates of overall decline in abundance and biomass of great concern. A major study in China (Zhou et al. 2023) assessed the abundance and trophic interactions from catches of insects over 18 years (2003–2020) by searchlight traps (3,124,778 individual insects) and radar observations of migrants (9,183,601 individuals), collectively perhaps the largest samples yet assessed in detail. A major aim of that study was to examine changes in food webs based on population trends of 98 airborne insect species and characterising their associated natural enemy–host interactions, comprising 124 such interactions. The natural enemy declines reported over this period (0.85%/year for a total of 15.24%) led multiple host species populations to become less stable. Zhou et al. found this indicating 'a critical decline in (insect) functional diversity and a steady loss in ecological resilience' and also illustrated 'the gravity of the regional (insect) biodiversity crisis'.

Two major difficulties persist: the lack of sound baseline information and of long-term monitoring studies from which such baselines may be derived, and lack of rigorous analyses that could account for unobserved events. Research on changes in insect diversity and abundance needs two complementary approaches: (1) to ascertain any directional change and, if possible, determine its magnitude and rate and (2) to seek correlation with possible causes and drivers (Lewinsohn et al. 2022). Especially in the tropics, foundation data on insect diversity and richness is sparse—but losses there may be far greater than elsewhere, simply because so many insects are present, and many of the losses will be of species that have not even been detected and can never be known. In the case of Singapore butterflies, Theng et al. (2020) built on previous studies, recent and historical species lists and direct examination of major institutional collections to evaluate extirpation rates since 1854 on this heavily urbanised but well-surveyed city-state island. From a database of 413 native butterfly species, 132 (32%) were recorded as extirpated. Adding the estimated further 104 unknown species extirpated before they were even discovered raised the level of loss to 46% (41–51%). Many extirpations took place more than a century ago, reflecting a period of large-scale destruction of primary forests and associated loss of forest specialist species. The second half of the twentieth century was characterised by rapid urbanisation in Singapore during which large areas of agricultural land were subsumed, and also linked with butterfly extirpations. This study drew attention to the likely more widespread importance of 'unknown extinctions'; these have only rarely been estimated but—as here—may substantially increase the true losses in insect faunas.

Study of museum and related historical collections in which specimens have been deposited over a considerable period may indeed provide suggestions of losses, but such data must be interpreted carefully and cautiously. In Australia, as elsewhere, the bulk of Lepidoptera in major collections is the outcome of hobbyist (broadly 'non-professional') activities, and various attempts have been made to assess incidence of species at a site 'then' and 'now'. Most such attempts cannot assess the completeness of early collections, but 'species presence' based on specimens is unambiguous. One of the most useful applications is for the rarest species, predominantly of the 'most collectable taxa' (Coleoptera, Lepidoptera), because in the era before collecting became a social stigma or was prohibited or regulated, these may have been sought actively as cabinet specimens and are 'over-represented' in collections. Some of those species are among those now given priority for conservation interest, and early records can help to document changes in range. One caveat, mentioned in some historical essays, is that occasionally collectors might falsify data labels for such species to protect 'their' special localities from exploitation by others! In contrast, many common species are not as easy to appraise in detail, because they tend to be overlooked by collectors and not accumulated comprehensively. Incidence of insect specimens, whether in field samples or museum collections, can be rather serendipitous and unpredictable, and in many cases their declines or losses must be inferred rather than proven.

The complex associations between (1) insects and plants or (2) insects and other animals are often very specific, so the loss of either party may harm the other. High plant diversity is commonly associated with high insect diversity. Many different categories of interactions are involved, and unified by the partners being specialised and, in many cases, interdependent. Specific insect pollinators are needed by many plants, for example and the decline of a mammal, bird, or other hosts can affect a specific parasite. In both contexts, that specificity means that few or no alternative associations are possible. Pollinator declines have become a strong focus in insect conservation as human food supply becomes less assured, but other plant species are inevitably involved as well. For a long time, the classic case of this coextinction was considered to be loss of the previously abundant Passenger pigeon (*Ectopistes migratorius*) in North America, and consequent extinction of its two supposedly specific louse parasites; both these have now been found to exist on other hosts. A more recent concern (Kehoe et al. 2020) is that declines of both the black and white rhinoceroses may increase vulnerability of a specialist parasite (the Stomach bot-fly, *Gyrostigma rhinocerontis*, Oestridae). Also writing on oestrids, Colwell et al. (2009) pointed out that extinctions of species associated with threatened mammal hosts may pass unnoticed through 'benign neglect' in that they are so poorly known. That sentiment applies very widely across many insect taxa.

Extending from such definable species–species interactions, species in communities are connected more intricately in complex interaction webs, as exemplified by Zhou et al. (2023, above). Loss of a single insect species may thus induce much broader ecological impacts, and losses or changes in functional guilds affect ecological processes (Chap. 4). Declines of necrophagous insects, mostly Coleoptera and Diptera and not yet assessed but possibly susceptible to climate change, could hypothetically lead to delayed colonisation of corpses and changed patterns of succession due to different responses between necrophagous species (Amendt 2021).

Reported declines vary greatly in extent across time, space, and the focal taxonomic insect group, but several studies note annual declines in abundance of around 1–2%—a figure that could sum to >10%/decade: as Wagner et al. (2021) noted, this suggestion provides ample cause for concern.

Whatever their real extent, recent and continuing declines of insects are a serious and formidable environmental problem with wide implications for the future of

Earth's biodiversity and ecological functions. Zhou et al. (2023) councelled that the changes they documented in China implied that active conservation of insect communities is pivotal to ensuring human food security as global environments change. Many of the reported changes can be linked directly with anthropogenic activities, impacting insects in many ways. Thus, despite the numerous uncertainties over incidence, extent, and details of insect declines, we cannot afford to wait for proof of all cases before acting to conserve them.

Communicating this uncertainty convincingly requires skillful and persistent effort but, as Saunders et al. (2020) wrote, 'We have not yet lost all insects, and there is still time to protect global insect diversity if we act now'. And, from Forister et al. (2019) 'We know enough to act now', commenting also that if we await full understanding of the mechanisms linking drivers to declines, we risk losing a good number of our subjects along the way. The expression of urgency needed for insect conservation is, indeed, widespread and more critical study of declines is a significant component of advance, in helping to understand their causes and the factors we need to address. The media attention from reported insect declines—whatever inaccuracies they promote—is also an opportunity to improve public understanding of insect life, and to debate the issues involved. Because many insect declines directly affect human welfare as well as ecological function and integrity, understanding their causes and countering them has considerable—and vital—practical and political importance for both humanity and the natural world.

References

Amendt J (2021) Insect decline—a forensic issue? Insects 12(4):324. https://doi.org/10.3390/insects12040324

Bluthgen N, Staab M, Achury R, Weisser WW (2022) Unravelling insect declines: can space replace time? Biol Lett 18:20210666. https://doi.org/10.1098/rsbl.20-21/0666

Colwell DD, Otranto D, Stevens JR (2009) Oestrid flies: eradication and extinction versus biodiversity. Trends Parasitol 25:500–504. https://doi.org/10.1016/j.pt.2009.07.011

Conrad KF, Warren MS, Fox R, Parsons MS, Woiwod IP (2006) Rapid declines of common, widespread British moths provide evidence of an insect biodiversity crisis. Biol Conserv 132:279–291. https://doi.org/10.1016/j.biocon.2006.04.20

Dicks LV, Grames E, Bowler DE, Isaac NJB (2024) Insect declines – an overview of current knowledge on the status of the world's insects. In: Pryke JS, Samways MJ, New TR, Cardoso P, Gaigher R (eds) Routledge handbook of insect conservation. Routledge, London and New York, pp 77–91

Didham RK, Basset Y, Collins CM, Leather SR, Littlewood NA (et al.) (2020) Interpreting insect declines: seven challenges and a way forward. Ins Conserv Divers 13: 103–114. https://doi.org/10.1111/icad.12408

Dunn RR (2005) Modern insect extinctions, the neglected majority. Conserv Biol 19:1030–1036. https://doi.org/10.1111/j.1523-1739.2005/00078.x

Forister ML, Pelton EM, Black SH (2019) Declines in insect abundance and diversity: we know enough to act now. Conserv Sci Pract 2019:e80. https://doi.org/10.1111/csp2.80

Fox R (2013) The decline of moths in Great Britain: a review of possible causes. Insect Conserv Divers 6:5–19

Fox R, Oliver TH, Harrower C, Parsons MS, Thomas CD, Roy DB (2014) Long-term changes to the frequency of occurrence of British moths are consistent with opposing and synergistic effects of climate and land-use changes. J Appl Ecol 51:949–957. https://doi.org/10.1111/1365-2664.12256

Fox R, Dennis EB, Harrower CA, Blumgart D, Bell JR (et al.) (2021) The state of Britain's larger moths 2021. Butterfly Conservation, Rothamsted Research and UK Centre for Ecology and Hydrology. Wareham, Dorset

Gossner MM, Menzel F, Simons NK (2023) Less overall, but more of the same: drivers of insect population trends lead to community homogenization. Biol Lett 19:20230007. https://doi.org/10.1098/rsbl.2023.0007

Goulson D (2019) The insect apocalypse, and why it matters. Curr Biol 9:R967–R971

Goulson D (2021) Silent earth. Averting the insect apocalypse. Jonathan Cape, London

Hallmann CA, Sorg M, Jongejans E, Siepel H, Hofland N (et al.) (2017) More than 75 percent decline over 27 years in total flying insect biomass in protected areas. PLoS One 12:e0185809. https://doi.org/10.1371/journal.pone.0185809

Janzen DH, Hallwachs W (2019) Perspective: where might be many tropical insects? Biol Conserv 233:102–108. https://doi.org/10.1016/j.biocon.2019.02.030

Kehoe R, Frago E, Sanders D (2020) Cascading extinctions as a hidden driver of insect decline. Ecol Entomol 46(4):743–756. https://doi.org/10.1111/een.12985

Leather SR (2019) "Insectageddon"—bigger headlines, more hype, but where's the funding? https://simonleather.wordpress.com/2019/02/15/insectageddon

Lewinsohn TM, Agostini K, Freitas AVL, Melo SAS (2022) Insect decline in Brazil: an appraisal of current evidence. Biol Lett 18:20220219. https://doi.org/10.1098/rsbl.2022.02191

Lewis OT, Basset Y (2007) Insect conservation in tropical forests. In: Stewart AJA, New TR, Lewis OT (eds) Insect conservation biology. CAB International, Wallingford, pp 34–56

Lister BC, Garcia A (2018) Climate-driven declines in arthropod abundance restructure a rainforest food web. Proc Nat Acad Sci 115(44):E10397–E10406. https://doi.org/10.1073/PNAS.1722477115

Lockwood JA (2004) Locust. The devastating rise and mysterious disappearance of the insect that shaped the American frontier. Basic Books, New York

Miller JC, Hammond PC, Ross DNR (2003) Distribution and functional roles of rare and uncommon moths (Lepidoptera: Noctuidae: Plusiinae) across a coniferous forest landscape. Ann Entomol Soc Am 96:847–855

Milman O (2022) The insect crisis. The fall of the tiny empires that run the world. Atlantic Books, London

New TR (2011) Butterfly conservation in South-Eastern Australia: progress and prospects. Springer, Dordrecht

Pyle RM (2012) The origins and history of insect conservation in the United States. In: New TR (ed) Insect conservation: past, present and prospects. Springer, Dordrecht, pp 157–170

Sanchez-Bayo F, Wyckhuys KAG (2019) Worldwide decline of the entomofauna: a review of its drivers. Biol Conserv 232:8–27. https://doi.org/10.1016/j.biocon.2019.01.020

Sanchez-Bayo F, Wyckhuys KAG (2021) Further evidence for a global decline of the entomofauna. Austr Entomol 60:9–26. https://doi.org/10.1111/aen.12509

Saunders ME, Janes JK, O'Hanlon JC (2020) Moving on from the insect apocalypse narrative: engaging with evidence-based insect conservation. Bioscience 70(1):80–89. https://doi.org/10.1003/biosci/biz143

Schowalter TD, Pandey M, Presley SJ, Willig MR, Zimmerman JK (2021) Arthropods are not declining but are responsive to disturbance in the Luquillo experimental Forest, Puerto Rico. Proc Nat Acad Sci 118(2):e2002556117. https://doi.org/10.1073/pnas.2002556117

Theng M, Jusoh WFA, Jain A, Huertas B, Tan DJX (et al.) (2020) A comprehensives assessment of diversity loss in a well-documented tropical insect fauna: almost half of Singapore's butterfly species extirpated in 160 years. Biol Conserv 242 (2020) 108401. https://doi.org/10.1016/j.biocon.2019.108401

Thomas CD, Hefin-Jones T, Hartley SE (2019) "Insectageddon": a call for more robust data and rigorous analysis. Glob Change Biol 25:1891–1892. https://doi.org/10.1111/gcb.14608

van Klink R, Bowler DE, Gongalsky KB, Swengel AB, Gentile A, Chase JM (2020) Meta-analysis reveals declines in terrestrial but increase in freshwater insect abundances. Science 368:437–420

van Strien AJ, van Swaay CAM, van Strien-van LWTFH, Poot MJM, WallisDeVries MF (2019) Over a century of data reveal more than 80% decline in butterflies in The Netherlands. Biol Conserv 234:116–122. https://doi.org/10.1016/j.biocon.2019.03.023

Wagner DL (2020) Insect declines in the Anthropocene. Annu Rev Entomol 65:457–480. https://doi.org/10.1146/annurev-ento-011019-025151

Wagner DL, Fox R, Salcido DM, Dyer LA (2021) A window to the world of global insect declines: moth biodiversity trends are complex and heterogeneous. Proc Natl Acad Sci 118(2):e2002549117

Zhou Y, Zhang H, Liu D, Khashaveh A, Li Q (et al.) (2023) Long-term insect censuses capture progressive loss of ecosystem functioning in East Asia. Sci Adv 9, eade9241

Chapter 3
Causes of Concern: The Main Threats to Insects

3.1 Introduction

The reality of many insect declines, and widely accepted likelihood that these may be more widespread than documentation implies and will accelerate in the near future, necessitates determining their causes, and finding how these might be ameliorated. Entomologists recognise about five main anthropogenic 'drivers' of insect declines, all complex and actually or potentially interacting in many ways and often unpredictably. This framework is related to the wider threat classification scheme defined initially by Salafsky et al. (2008) and used in IUCN Red List species assessments (p. 112) to describe the activities or processes that impair the survival of species, as summarised in Table 3.1. Co-occurring disturbances may thus be synergistic, antagonistic, or independent if they interact, and their effects are difficult to compartmentalise. The overview for Australia compiled by Yen and Butcher (1997) contains much early background to concerns.

Although terminology varies widely, those five major drivers encompass (1) changes to natural habitats, including loss, degradation, and fragmentation, from a range of activities including urbanisation, industrial and infrastructure developments, agriculture, forestry, and mining; (2) pollution, most commonly through chemicals such as pesticides and fertilisers, but with light pollution an increasing concern; (3) invasive alien species, deliberately introduced or accidentally arriving in areas beyond their natural range, and interacting with species in their 'new' environment; (4) overexploitation, for food, commercial collecting interests and other purposes; and (5) climate changes, less tangible but overlying and influencing the impacts and extent of all the above. Interactions between these drivers can be complex, synergistic, or antagonistic, in many cases poorly understood, inconsistent, and are often unpredictable in detail. It can be very difficult to determine which of many interacting stressors is or are actually the most significant. Forister et al. (2019) used the analogy that insects face a 'firing squad' of threats, and it may not

T. R. New, *Insect Conservation in Australia: Why and How*,
https://doi.org/10.1007/978-3-031-66631-5_3

Table 3.1 The major categories of direct threats to biodiversity, as the World Conservation Union—Conservation Measures Partnership classification (version1.1). Abbreviated to major headings from Table 3.1 of Salafsky et al. (2008)

Threat category	Definition
Residential and commercial development	Human settlements or other non-agricultural land uses with a substantial footprint
Agriculture and aquaculture	Threats from farming and ranching as a result of agricultural expansion and intensification, including silviculture, mariculture, and aquaculture
Energy production and mining	Threats from production of non-biological resources
Transportation and service corridors	Threats from long narrow transport corridors and the vehicles that use them including associated wildlife mortality
Biological resource use	Threats from consumptive use of 'wild' biological resources including deliberate and unintentional harvesting effects; also persecution or control of specific species
Human intrusions and disturbance	Threats from human activities that alter, destroy, and disturb habitats and species associated with non-consumptive uses of biological resources
Natural system modifications	Threats from actions that convert or degrade habitats in service of 'managing' natural or semi-natural systems, often to improve human welfare
Invasive and other problematic species and genes	Threats from non-native and native plant, animals, pathogens/microbes, or genetic materials that have or are predicted to have harmful effects on biodiversity following their introduction, spread, and/or increase in abundance
Pollution	Threats from introduction of exotic and/or excess materials or energy from point and nonpoint sources
Geological events	Threats from catastrophic geological events
Climate change and severe weather	Long-term climatic changes that may be linked to global warming and other severe climatic or weather events outside the natural range of variation that could wipe out a vulnerable species or habitat

be clear which is the 'killing shot'; however, 'we know that bullets are flying and where they are coming from, and even when the squad is unchanged a bullet from another source may be the mortal hit'.

Two different spatial scales of disturbance are sometimes recognised (Wagner et al. 2021), as (1) large scale, such as continental, largely reflecting land-use changes and climate influences, and (2) site-specific threats, such as local clearing or pollution. The latter are commonly the initial concerns, as tangible and of concern to a local human community seeking to conserve their natural environment.

Each driver may have direct immediate impacts on native insect species and communities by killing them (as well as affecting development, physiology and/or behaviour and ecological interactions), but also cause longer-term changes as interactions and conditions for the communities are modified and, in some cases, key components lost. The insects in any community are interlinked through complex networks of interactions, so that losses of even a single species may have far-reaching consequences. In essence, impacts of the initial visible disturbance may be multiplied as its ramifications become apparent, and the details of such outcomes are usually unpredictable. One difficult-to-study aspect is the development of 'cascading extinctions', in which single extinctions can induce waves of secondary extinctions. Kehoe et al. (2020) discussed how each of the five above drivers can lead to cascading extinctions, but argued that habitat loss and pollution may be the most influential because they directly change community structure, the physical environment and community robustness. As they claimed, most direct impacts of anthropogenic drivers can lead to coextinctions and cascading extinctions—together 'secondary extinctions'—as a major component of insect declines.

Environments change, in many cases leading to them becoming more or less hospitable to their present inhabitants. Change may be natural—patterns of vegetation succession transform areas in a largely predictable way, and at rates to which insect inhabitants have become adapted. In contrast, imposed anthropogenic stressors may act rapidly and directly threaten insect survival. The major, and universal, measurable changes to insect environments are the loss, despoliation, and isolation of 'habitats' through a variety of stressors or 'threats'. Most obviously, loss of natural vegetation systems—such as clearing of tropical forests and destruction of native grasslands—and draining of wetlands for purposes such as agricultural use, urban and industrial development continue largely unabated and, in many parts of the world are difficult to monitor, regulate, or oppose. These habitats also support a panoply of 'visible' insects that can help draw attention to their values. The less obvious, but still diverse, insect fauna of soils and ground litter is generally overlooked and plays little part in insect conservation despite its ecological importance in nutrient cycling and soil health—and its vulnerability as soils are tilled, compacted, chemically polluted, and poisoned over large areas. Declines in soil health are a global problem that affects stability of many agricultural systems, and 'healthy soils' are 'the foundation for complex biological communities that promote diversity of interacting organisms' (Alyokhin et al. 2020), with maintenance of soil quality also contributing to improved pest management. Soil erosion, increasing with land clearance, is also associated with insect habitat loss.

Many changes to insect habitats may be severe—the clearing of forest or draining of a wetland or waterbody essentially removes abruptly the resources needed by myriad insect species. Floods and fire are both key disturbances to ecological systems, with impacts reflecting intensity and extent as well as season of incidence. Selective felling, creating gradients of tree density and perhaps reducing selected species, affects both the spectrum of insect species present and the changing distribution of species adapted to dense forests and more open forest environments. As one of rather few detailed examples, Koh et al. (2004) attributed the loss of a number of butterfly species in Singapore to loss of their host plants: loss of 208 species of plants used by specialist butterflies was associated with coextinction of at least 56 butterfly species. Losses of tropical rainforests in the neotropics and southeast Asia receive considerable international publicity and concern, but the fate of many affected areas and loss of less conspicuous vegetation systems is little advertised and their extent is unappreciated. Janzen (in Janzen and Hallwachs 2019) noted that by 1990, 49% of Costa Rica was cattle pasture, much of it derived from formerly forested areas. Vegetation structure provides the architecture on which utility resources (below) are founded, whilst vegetation species furnish the basis for most consumables—either directly (for herbivores) or as the food for the prey or hosts of higher-level consumers. Presence of 'rare' plant species may indicate presence of equally restricted insect species, and many rare plants are clustered in small isolated habitat patches produced by landscape features as 'ecological islands' (Cartright 2019).

The full ramifications of forest disturbance on insect life can be far more subtle and varied than simply 'habitat destruction'. The variety of possible or hypothesised impacts on bumblebees noted by Mola et al. (2021) (Table 3.2), for example is likely to be mirrored by many other insects, and in other habitats. Some such suggested impacts on other insects are simply conjecture, but collectively indicate the range of possible effects that merit consideration. Likewise, some impacts may initially appear to be insignificant—but only careful assessment can confirm this.

The term 'habitat' has traditionally been equated with 'a place to live' and this use corresponds with widespread recognition of insects associated with particular ecosystems, such as 'forest insects', 'grassland insects', 'freshwater insects', and many similar divisions that reflect the full variety of such systems, each then a

Table 3.2 Summary of threats to forests and their potential impacts on bumblebee populations (from Mola et al. 2021, in which references to all hypothesised impacts are included)

Threat	Hypothesised impact(s) on bumblebees
Fragmentation	Loss of forest habitat specialists, increase in generalist species; change in edge microclimate affecting foraging, nesting, and overwintering conditions
Loss of old-growth forests	Changes in forest floor structure suitable for overwintering; loss of understorey herbs; loss of old trees, stumps, and nesting cavities
Overbrowsing by deer	Loss of bumblebee forage plants; change in forest structure may affect suitability of nesting and overwintering, directionality unknown
Introduced earthworms	Changes to forest floor structure, moisture, and soil compaction may affect overwintering and nesting; loss of bumblebee forage plants
Wild and prescribed fire	Varied impacts depending on forest type, presumed increase in floral abundance due to increased light levels and postfire bloom; potential mortality of queens and colonies during overwintering or nesting; loss of microclimate buffering if canopy severely reduced
Logging	Varied impacts depending on logging intensity, type of machinery used, seasonality and disturbance etc., likely increase in forage and bee abundance, especially along edges; potential long-term negative impact due to loss of microhabitat structure
Invasive plants	Loss of floral abundance although some invaders are suitable forage; increased shade reduces foraging
Changed flowering phenology	Phenological mismatch
Pesticide concentration	Potential transfer to overwintering queens in soil; uptake into nectar and pollen

'habitat'. Each such category is very broad, and can be subdivided to reflect far more specialised occupancy—so that many categories of each of the above support different insects with more specialised needs. Some ecologists now prefer to recognise a 'resource-based' concept of habitat in which a simple 'place' gives way to an environment in which a species' critical resources are present and accessible. Dennis et al. (2006) and Dennis (2010) designated three categories of resources as (1) consumables, the specific foods needed by all feeding stages and including larval food plants or prey, adult supplies of pollen, nectar, plants or prey, and any obligatory associates such as mutualists or symbionts; (2) utilities, the environmental features that enable the species to undertake its full range of normal development and behaviour: they may include basking sites, territorial posts, overwintering, aestivating, oviposition and pupation sites, enemy-free space, and others which often cannot be defined fully for any given species but essentially endorse the need for environmental heterogeneity; and (3) conditioners, affecting how those resources may be gained and used: climate conditions and disturbance regimes can fall into this category of influence.

All these parameters are dynamic, changing continually and with their interactions also changing. Natural successions change the variety of plants present and, so, the insect herbivores that might occur; a given resource may be available for only a limited season; interactions may change availability and supply; and disturbance of any sort may affect resources. For any insect species, critical resources may differ across its distribution—many herbivores exploit different food plants in different places, and climatic differences may affect developmental patterns and seasonal activity.

Nevertheless, evaluating impacts of environmental changes and determining 'threats' is the major basis for assessing conservation needs and planning management. Much insect conservation has focused mainly on individual sites where 'threatened' insect species occur, but a wider 'landscape-level' approach is increasingly advocated. This chapter is a brief overview of the major concerns from the drivers noted above, the major threats to insect life. Each of these factors has generated a vast and complex literature evaluating its impacts across different contexts and environments.

3.2 Habitat Loss and Change

Habitat change is regarded widely as the predominant cause of insect decline and loss, with three main components contributing to this. First, direct destruction of vegetation or water bodies eliminates the critical resources and environmental conditions on which many resident insects depend, and causes changes to the physical structure and climate of the environment. Second, partial or localised destruction can have similar effects even if changes are less severe: changed vegetation composition, for example affects the composition of insect communities and selective logging eliminates specialists that depend on the logged target tree species. Third, and reflecting the scale of destruction, the pattern of habitat availability over the wider landscape may be changed to leave remaining suitable habitat patches smaller and increasingly isolated, more difficult for insects to reach, and more susceptible to other threats such as invasion by alien species that further erode habitat quality. Lowered connectivity and availability of suitable habitat are almost universal concerns; indeed, it has been claimed that habitat configuration in the landscape may be the single greatest threat to biodiversity. Insects must be able to reach those patches, and their quality must be sustained—as a general presumption, poor connectivity hampers range changes, and smaller patches support fewer species than larger patches, and are more susceptible to edge effects. Small remnant patches of vegetated habitat may entirely lack unchanged interior conditions. Additionally, small areas may be able to support only small populations of the resident species, perhaps increasing risk of their extirpation from inbreeding; the resources (such as particular plant species) needed by many specialist insect species may simply not be present. The broad effect of replacement of native vegetation and changed land use is simplification and increased homogeneity of the previously more diverse plant communities on which many insects depend. In many communities, insect diversity depends on plant richness and species composition, vegetation height and cover ('architecture'), and the plant functional groups represented; such biotic complexity and heterogeneity can foster insect diversity. The vast numbers of insect herbivores, and that many of them are wholly or largely host plant specific, dictates that the richness of native plants in an area may be a critical need. Priority conservation areas and reserves throughout the world need to embrace endemic plant richness as a major correlate of endemic insect herbivore survival (Fonseca 2009).

However, habitat loss is often less obvious than major vegetation changes. Numerous insects depend on dead wood, for example and removal of dead trees or fallen branches or logs be harmful; leaf litter insects are lost if litter is burned or otherwise removed; mowing or slashing grasses or other low vegetation can remove critical food resources, including nectar sources; tilling may affect soil insects; introductions of alien plants can create novel interactions, some leading to ecological traps (p. 47); and clearing of riparian vegetation—as with many local disturbances—can change microclimates on which particular insects depend. In short, almost any disturbance may have unforeseen impacts on insect life.

The major drivers of habitat change are noted below. All have been discussed exhaustively in recent years, but additional ramifications continue to be documented. Land-use change, inexorably linked with losses of floral diversity and extent and wider losses of habitat structural complexity, is a major driver of insect declines, as demonstrated by changes associated with urban and agricultural developments. Urbanisation and agricultural intensification generate two major categories of transformative anthropogenic landscapes, based on essentially artificial features far different from the environments they replace, and providing conditions inimicable to wellbeing of many earlier residents. The conservation measures noted for insects in these extremely changed environments (see also Chap. 7) have far wider benefits to biodiversity, in providing/restoring resources needed by birds, mammals, and other more 'charismatic' taxa as well as hordes of other invertebrates.

3.2.1 Urbanisation

Very broadly, 'urbanisation' encompasses human modifications of landscapes for settlement and associated functions such as industrial and commercial services, in essence the development of towns and cities but with very varied scales and impacts that may extend far beyond their source areas. Some writers have referred to urban areas as 'hardscapes', in contrast to 'greenscapes' (Deacon and Samways 2021). Replacement of natural vegetation by sealed surfaces of concrete and tarmac is an extreme and abrupt loss of natural habitat, with the corollary that changed patterns of water run-off can rapidly affect the receiving water courses and their inhabitants. Urbanisation places immense pressure on freshwater aquatic environments, in addition to their direct loss. Increased need for human domiciles and related transport and support systems continue to envelop more land, and periurban areas are inexorably absorbed into development as new suburbs and settlements are established; the amount of land needed for urban developments and associated industrial, recreational, and transport amenities continues to burgeon and increase the 'built environment'. Ideally, that environment should promote human wellbeing and urban beauty whilst also maximising potential for conserving insects (Hunter and Hunter 2008). Both aquatic and terrestrial environments are candidates for that endeavour.

About half the world's population currently lives in urban environments, and this proportion is forecast widely to increase over the next few decades. Impacts are widespread, as increased road networks link urban centres, and roadside vegetation remnants can become important refuges for native insects, together with such patches within cities and towns. Indeed, any such remnants may need strenuous efforts to preserve them, whilst others (such as golf course 'rough' areas) also have conservation value. Not least, larger open areas commonly retain larger trees, with their associated herbivore and saproxylic insects. Many of the latter, indeed, depend on older trees with dead wood or cavities (Horak 2018), and 'urban forests' provide critical resources for many insects. Needs for protection of the oldest trees, sometimes dubbed 'veteran trees', and the availability of fallen woody debris are easily overlooked as developments occur. Indeed, so-called 'urban forests' (whether of remnant or planted trees) provide some of the most structurally complex insect habitats in those areas; urban forests comprise an upper canopy, one or more understorey layers, a ground cover layer, living and dead wood, leaf litter and an enriched soil layer, collectively a diverse array of invertebrate habitats. Their diverse and complex interactions with invertebrates render their conservation critical in fostering urban ecosystems that are resilient to increasing anthropogenic pressures (Kotze et al. 2022).

Much urban development occurs on land that has been cleared earlier for other purposes and is already changed considerably. Developments that involve clearing of pristine vegetation may have greater conservation significance and even tiny remaining patches or enclaves of natural vegetation may need strenuous protection as island refugia for species eliminated from the surrounding landscape, and/or as stepping stones facilitating connectivity (p. 135). Small patches of urban unmanaged vegetation are often regarded as 'wastelands' (Labadessa and Ancillotto 2023) perhaps left accidentally by developmental planning and regarded as unimportant. However, they are key to conserving biodiversity in many urban areas, and undisturbed urban patches have been described as 'bioarks' (Diamond et al. 2023); many insects can continue to thrive in small vegetated areas that would not generally be considered important in conservation. Urban green spaces are acknowledged widely as important for native bee communities and pollinator maintenance in urbanised landscapes in Australia (Threlfall et al. 2015). They also host other insects and may be suitable as introduction site translocations (p. 124).

Less tangible impacts on insects from the 'urban heat island effect' influencing development rates and artificial lighting at night (popularly 'ALAN', p. 47) are also widely claimed.

Inner urban developments are far different from the systems they replace, and continue to be alienated as density of construction increases—such as through former home gardens being subsumed for 'second dwellings' or house extensions. Natural vegetation can be sparse, and replaced by alien plant species or low diversity associations such as intensively managed grass lawns; open areas are limited, and isolated other than by roadways; sanitation measures (such as weed removal and mowing) that oppose increase of plant diversity are common; and urban areas are associated with many forms of pollution. Loss of native vegetation is a key driver of decline of insect herbivores both in temperate regions and the tropics.

The 'urban estate' can be divided broadly into two major categories, within a complex mosaic of land uses and habitats, as (1) public open spaces managed by local authority, and ranging from parks and sports grounds (with larger open areas such as golf courses providing significant conservation opportunities) to roadsides strips and medians, and (2) private open spaces, namely home and other private gardens. Additional open spaces include (as examples) school grounds, cemeteries, community gardens, and constructs such as green roofs and green walls, all of which are receiving increased attention as conservation vehicles and using local or native plant species wherever possible. Widespread recommendations for creating more green spaces in cities, focusing on increasing the variety and diversity of native plants, are a key management step. Collectively, these open urban areas provide abundant opportunities for sympathetic management for insects (New 2015), not least because their conditions and treatments can be manipulated easily, from affecting floristic composition and pesticide/fertiliser applications to management such as frequency, intensity, and seasonality of mowing, use of mulching (with the cut vegetation left on site), weeding or slashing, and restoration/rehabilitation by selective plantings or wild seed mixtures. However, public perceptions of the urban landscape vary enormously and coordination between landscape architects, urban planners, and entomologists is needed to improve public support for insect conservation in urban areas (Hunter and Hunter 2008). Their comment that 'It is at best naïve, and at worst unethical, to ignore public perception of the urban landscape in conservation management' remains apt. A survey of perceptions by children of their schoolyards (as the open area with which they are most likely to be familiar and occupy, and assessed by seeking opinions of computer-enhanced photographs) showed that areas enriched with vegetation were preferred over traditional grounds with paved/asphalted areas and grassed sections (Lindemann-Matthies and Kohler 2019): in this case, small-scale biodiverse-rich enrichments increased appeal of schoolyards, in conjunction with functional elements such as gravel paths and benches for relaxation. Tacitly embracing such sentiments, the five principles of 'Biodiversity Sensitive Urban Design' advocated by Garrard et al. (2018) apply well to insect needs, as (1) maintain and introduce habitat; (2) facilitate dispersal; (3) minimise threats and anthropogenic disturbance; (4) facilitate natural ecological processes; and (5) improve potential for positive human–nature interactions. The combination of needs for 'preservation' and 'restoration' of urban open spaces is acknowledged increasingly.

Home gardening is one of the most common activities whereby people 'connect with nature' (Vanderstock et al. 2022) and measures such as installing 'bee hotels' to provide nesting sites (which, however, can also harbour alien species and perhaps facilitate their spread: Geslin et al. 2020) and planting local native species suitable as nectar and pollen sources or larval food plants are valuable conservation steps in replacing or augmenting resources. Some urban plantings have been used as 'pollinometers' in surveys of insect pollinators, by recording their visitors over standard periods: in studies of pollinators in cities in Germany, potted Red clover (*Trifolium pratense*) plants were deployed for this purpose (Theodorou et al. 2020). 'Butterfly gardening' (p. 133) is increasingly widespread, and the principles—enhancing floral resources for nectar and encouraging larval food plants—benefit numerous less conspicuous insects. Despite frequent recommendations to use only native flora in such exercises, alien species may be useful in contexts such as extending the flowering period and increasing seasonal supply of resources for insects. The aesthetic and recreational attractiveness of mown lawns as amenities in home gardens and parks is offset by many being monocultures, and recommendations have been made repeatedly to increase plant variety in lawns and, even, to reduce their extent in favour of other plantings. The mosaic of home gardens inevitably includes many tradeoffs between aesthetic preference and suitability for conservation; one role of urban planning is to balance the quality of public open spaces between conservation and ecosystem benefits and 'disservices' as perceived by people. The latter may include non-sanitised areas, dead wood, weeds, long grass, and other features that contribute resources and heterogeneity.

The putative functional isolation of many inner urban open spaces can in principle be reduced by promoting greater connectivity through improving quality of roadside areas including greenways, nature strips, and road medians to include native vegetation that can support native insects and facilitate their wider dispersal. As with many aspects of the built environment, their presence can become somewhat unpredictable, as planning priorities change, renovations, additional constructions or roadworks occur and previously unused open spaces become committed to development. Campaigns to conserve insects found in urban spaces or periurban areas committed to development of new suburbs or industrial development can become protracted and polarised.

Many studies of insects along 'urban gradients', from rural areas to city or town centres have implied both loss of richness and losses of specialist species with increased urbanisation, and that inner urban faunas are dominated by generalist species and higher levels of alien species—some associated with the large variety of alien plants found in home gardens and parks. Complexes of native natural enemies change in parallel. Effects are complex, making generalisations uncertain: (1) reduced numbers of rare bee species were associated with urban and agricultural landscapes in the United States, with these landscapes presumed to negatively affect rare (specialist) bees (Harrison et al. 2019); (2) the density of domestic honeybee colonies in Paris, France, negatively affected flower visitation rates by some native pollinators and whilst honeybees tended to be most active on managed flowers, wild bees equally visited managed and wild species (Ropars et al. 2019); and (3) surveys in Germany of flying insects showed that flies and butterflies tended to have lower richness and pollination rates in cities but bees had the opposite trend (Theodorou et al. 2020). However, studies in many parts of the world have shown that pollinators, including native bees can remain abundant in city open spaces and often include species absent from the surrounding landscape (Hall et al. 2017), implying that the variety of foraging and nesting areas within cities renders them a 'refuge' for insect pollinators and emphasises their conservation values. As Hall et al. commented, intensifying conservation efforts for urban insect pollinators creates opportunities for planning that acknowledges the ecological complexity of urban environments and garners interest from many sources.

Practical insect conservation must heed the reality that, whilst 'megacities' have major impacts on insect biodiversity, urban impacts occur at much lower levels and scales of urbanisation—such as within informal settlements and suburban locations (Diamond et al. 2023). But, as a general principle, urban insect conservation demands the twin approaches of (1) reducing exposure to threats and (2) enhancing the resources they need. For urban forest invertebrates, Kotze et al. (2022) nominated resource enhancement to include supply of coarse woody debris and dead wood, areas of exposed soil, water bodies, native plant species, managing vegetation structure, and ensuring resources are available for all life stages. They also specified reduced exposure to urban heat island effects, ALAN, air and water pollution, natural disasters, overabundant species, and human neglect—empowering people to become 'stewards' for wildlife is a key remedial step. Any individual situation will necessitate variations from this protocol, but the indicated range of components indicates the complexity of any such exercise. Native insects subjected to urbanisation or other major habitat changes may either persist as tolerating the new environment, or become extinct, as non-tolerant; the balance between these trends shapes the new resident community. Essentially, the urban environment 'filters' the local species pool to allow those with particular traits to persist; thus, for Orthoptera in Rome, local extinction rates were higher for species with low mobility and fertility (numbers of eggs produced) and that had narrower (more specialised) climatic demands (Ancillotto and Labadessa 2023).

3.2.2 Agriculture

Development of intensive agriculture for human food production over vast land areas is one of the greatest terrestrial transformative processes, by which natural environments are changed to simplified ecosystems by land-clearing, to leave areas often characterised by large-scale monocultures of alien plant species such as food crops or exotic grasses for 'improved pasture' for livestock. Rates of clearing vegetation for agriculture have recently been described as 'unprecedented', and are likely to increase even further with intensification of agricultural production and needs to assure human food supplies.

A cleared agricultural landscape has been characterised by having less than 5% natural habitat in a 1-Km radius (Tscharntke et al. 2012), and intensification of agriculture during the middle twentieth century saw much traditional small-scale food production in Europe give way to large fields, by removal of boundary vegetation such as hedgerows to enable large areas in which larger machinery could operate. Enhancement and protection of crops by fertiliser and pesticide applications are widespread, and use of non-specific antihelmintics and other drenches for livestock can provide additional environmental impacts on native insects by reducing suitability and survival in animal wastes.

Historically, from around the middle of the twentieth century, intensive agriculture was associated with increased field size, achieved in Britain by removal of long-established field divides such as floristically rich hedgerows, removing much insect habitat and increasing distances between remaining natural patches. It is now acknowledged widely that such intensification also had some severe effects on the environment, largely because of habitat modifications and loss, non-target and accumulation effects of pesticides and fertilisers, and increased tillage. Considering the variety of impacts on pollinators, Tscharntke (2021) listed intensified agrochemical use, reduced crop diversity, crop monocultures replacing polycultures, field enlargements, losses of habitat patches supplying food and nesting sites, and 'overall homogenisation of landscape structure'.

More recently, in part as a response to integrated pest management involving naturally occurring predators and parasitoids of pest insects, restoring the hospitality of field margins to sustain these agents, together with pollinators has facilitated aspects of natural pest management, with accompanying reductions of insecticide use. 'Conservation biological control', sustaining naturally occurring predators and parasitoids of pest insects in agroecosystems is adopted widely in pest management, as a key tactic in reducing pesticide applications (p. 46) and as a keystone of 'agroecology'—this basically being the integration of ecological, economic, and social dimensions of the entire food production system (Rusch et al. 2017).

Much of the emphasis on insect conservation in agricultural ecosystems devolves on reversing the impacts of intensification, through restoring traditional farming systems and emphasising the importance of heterogeneity and providing insect habitats and resources (such as by more natural field margins with native vegetation, conserving and re-establishing hedges and other field divides, reducing field size, instituting mosaic cropping to replace large continuous areas of monocultures, and promoting connectivity throughout the landscape). These enhancements are accompanied by progressive reduction of harmful pesticides and more precise applications to avoid non-target effects. Many modifications, indeed, are motivated by the need to reduce pest impacts, through measures such as enhancing natural enemy populations and removal of pest resources. The key themes for insect conservation in agroecosystems devolve on ecological understanding, and that many native insects may be beneficial and able to thrive with little loss of economic productivity.

3.2.3 Forestry

Forest clearance is a major component of habitat change throughout the world. Many concerns over forest exploitation devolve on removal of trees for timber, woodchips, or other uses, as direct habitat destruction, extending to clear felling for alternative land use. Removal of fallen timber and dead wood also destroys the resources of numerous saproxylic insects and others using the wood as a refuge. 'Selective logging' appears less damaging than felling of entire forest stands, but effects such as soil compaction from heavy machinery, creation of forest gaps, or removal of particular tree species from mixed stands may all constitute threats by disturbing the local environment. Loss of old-growth forest for hardwood timber in Australia, an industry on which many livelihoods depend, is a threat of global significance in that—as for much tropical forest—those complex and ancient ecological associations are unlikely to ever be regenerated.

Commercial forestry, however, depends on managing and sustaining the tree crop. Much modern forestry involves single-species plantation establishment and maintenance, to produce uniform-aged monoculture stands of trees that can be harvested together. Initial preparations for plantation establishment involve clearing existing tree vegetation by felling and burning, with consequent destruction of the ground layer plants and other understorey, woody debris, and litter—all important insect resources—and changes to the soil characteristics. Plantation maintenance includes periodic suppression of undergrowth, thinning, and, where necessary, chemical control of pests (sometimes by ground- or aerial spraying), converting natural forest systems to homogenous vegetation in which understorey and other vegetation are suppressed. That sanitation also reduces (or wholly removes) the resources that can sustain the natural enemies of forest insect pests, and suppress their roles in conservation biological control.

Plantations are often of trees not native to the area of planting, so introducing and nurturing a predominant alien species—with potential for alien consumers to establish on them. Throughout the world, alien insects are amongst the most serious forestry pests, with non-target impacts on native insects from pest management a frequent concern.

In Australia, the main home-grown softwood source is through plantations of the alien Monterey pine (*Pinus radiata*), originally from North America but now very widespread, but a less appreciated and more subtle level of alien species is the planting of eucalypts native to one part of the country far from their natural range, and presenting an additional suite of interactions to which some native insects may respond. Many insects can to some extent change their habits to feed on alien trees. Others cannot, so that the Lepidoptera with larvae able to feed on *Pinus radiata* in Australia are a small subset of those present on the eucalypts the pines replace. Those moths and beetles feeding on eucalypts outside their natural Australian range may also lead to changed balance in local communities, but may also be susceptible to management.

Harmonising silvicultural operations with insect conservation to develop ecologically sound approaches to plantation management includes several 'needs', listed by Spagarino et al. (2001) as (1) quantifying the insect diversity present; (2) identifying and enumerating the insects present; (3) determining the vulnerable species; (4) studying their ecology and environment relationships; (5) validating the effectiveness and practicability of the new silvicultural methodologies; (6) applying these in production areas; and (7) where possible, enlarging studies to other taxa and relating these to changes observed amongst insects. Such long-term demands remain largely idealistic, but this list (devised for *Nothofagus* in Argentina) suggests parallel need for Australia. The first four items are the foundation for justifying insect conservation needs but are currently impossible to address comprehensively. Forest conservation remains a major national concern, and the need for better-informed and biodiversity-sensitive management in conjunction with increased security of forests in reserves are central issues for the future (New 2018). Until recently, forest insect conservation and forest pest management have been largely separate disciplines, and the recent trends towards creating forest landscapes in northern temperate regions that can reconcile these priorities by reducing severity of pest outbreaks and increasing security for insect diversity are increasing confidence that this may occur (Gazzea et al. 2024).

3.3 Pesticides and Pollution

Broadly, 'pollution' refers to environmental contamination by materials and processes that are in some way unnatural, usually with the implication that these may be harmful, often causing mortality. Pesticide and fertiliser applications are a mainstay of much productive agriculture, with intensification of agriculture leading to increased chemical treatments both to foster crop production and suppress the pests (many of them insects) that hamper that endeavour. Chemical pollution occurs in many forms, but a broader scope of 'pollution' now also includes light pollution and noise pollution, both of which can disrupt normal insect behaviour, with a variety of harmful effects. For any pollution, the range of non-lethal side-effects on insect development, reproduction, physiology, behaviour, and interactions remains largely speculative. Much pollution is a direct consequence of processes such as agricultural intensification and urbanisation.

The long-known undesirable non-target impacts of chemical pesticides, despite much regulation of applications, are a persistent concern for their impacts on non-target organisms and accumulation along food chains, as well as their 'drift' to other environments from aerial applications or transported as 'run-off' in water. The early concerns over bioaccumulation of DDT and others were the stimulus of far wider concerns over side-effects of chemical pesticides and, despite numerous advances in 'assuring safety', 'dose control', and 'control of applications', the vastly increased variety of chemicals devised for pest suppression (many with doubts over their non-target and other side-effects) lead to continuing concerns and search for alternatives. Impacts are clearly context-dependent, but in agriculture, forestry, domestic environments, and others concerns over the use of broad-spectrum pesticides—with their effects on non-target organisms and, perhaps, human health—have led to searches for alternatives and to more precise and well-considered applications.

As an especially widespread concern, the direct and indirect effects of pesticides on bees are especially well documented, with drift from agricultural areas to neighbouring natural environments a concern, and aspects of this still not understood clearly. Volatile herbicides, for example have been reported injuring plants up to 250 m from application sites, perhaps reducing foraging opportunities for bees (Mola et al. 2021). The loss of bee pollinators, related in part to their exposure to a 'cocktail' of many different chemicals (insecticides, herbicides, and fungicides), has been discussed extensively (Goulson 2019), with extensive political debate on the use of bee-killing pesticides. Recent concerns have emphasised, especially the roles of a class of neurotoxic insecticides called 'neonicotinoids', linked strongly with bee declines, but implicated also in losses of many others, including aquatic insects. Neonicotinoid use is now highly regulated (Wood and Goulson 2017; Hladik et al. 2018), as consequences of their use become more broadly acknowledged.

Increased applications of synthetic fertilisers, with changes to nitrogen regimes, are associated with changed floristic condition, composition, and botanical diversity (and, hence, with herbivorous insects and their natural enemies) on and near application sites, and pollution effects are extended by run-off into waterbodies, leading to eutrophication. Residues may also persist in soils, where impacts may occur on developing stages of some solitary bees and others (Mola et al. 2021). Other 'non-pesticide' chemical applications can also have poorly understood impacts. In North America, winter applications of salt (sodium chloride) to roads to de-ice them are widespread, with the result that soil salinity along roadsides is increased by run-off. This may lead to increased sodium content in floral nectar, perhaps increasing attractiveness to pollinators and increasing the ecological trap effect through elevating mortality (Lovett and Carr 2023). Similarly, pollution from vehicle engine combustion can lead to roadside plants having elevated levels of heavy metals and rendering them toxic or unpalatable to herbivores. Very broadly, this form of 'traffic pollution' can affect insect herbivores by (1) changing habitat quality; (2) changing host plant quality; and (3) affecting natural enemies by reducing or eliminating their impacts. Heavy metal accumulation may be especially significant (Karavin 2024).

'Light pollution' refers mainly to increased use of nocturnal lighting that disrupts insect behaviour. Light traps have long been a major collecting and monitoring tool in insect ecology and are based on the simple principle that many insects are attracted to a light source. The diverse ramifications of Artificial Lighting At Night ('ALAN') are increasing concerns as ALAN increases from urbanisation, communication networks, and industry and becomes more widespread. It was cited as a major contributor to moth declines in Britain (Fox 2013). Simple attraction of numerous phototactic insects from the surrounding area draws them from their natural habitats and provides a concentrated food supply for nocturnal predators (notably bats, but also birds and other insectivores), and may deplete populations of scarcer species. ALAN can thereby act as an 'ecological trap' removing individuals from a breeding population. However, a number of other impacts are involved (Boyes et al. 2021a, 2021b) and are difficult to interpret; overall impacts have only rarely been quantified, but the sobering figure cited by Peter et al. (2020) provokes concern over the fate of local faunas. They cited an earlier estimate of around 10^{11} moth deaths at light sources in Germany during a single summer.

More specialised concerns are exemplified by the use of cave lighting for tourists (p. 95) to facilitate viewing of Australian 'glowworms', the bioluminescent larvae of *Arachnocampa* spp. (Diptera: Keroplatidae, endemic to Australia and New Zealand). Exposure to artificial light can cause larvae to dim, leading to reduced attractiveness of the natural display to visitors and, perhaps, also reducing their capture success by not attracting prey. Both Driessen (2010) and Merritt and Clarke (2013) indicated how artificial light might be varied and monitored in caves to enhance glowworm conservation without diminishing the 'visitor experience'.

'Noise pollution' has been far less intensively studied but concerns that anthropogenic noise—such as from transportation, construction, and energy generation—can interfere with normal acoustic cues and have impacts on insects and arachnids (Classen-Rodriguez et al. 2021) suggested that sound affects species' behaviour and may lead to changed interactions in a community. Claasen-Rodriguez et al. distinguished three mechanisms by which behavioural alterations might be induced, as 'masking' (studied most in the context of mate detection in Orthoptera, where male stridulation is a frequent mating cue), 'distraction' (where attention is shifted from a target source to a noise source, perhaps affecting prey capture or foraging efficiency), and 'misleading' (where an anthropogenic noise is detected as a natural cue but triggers a response that is maladaptive). Both airborne and substrate-borne vibrations are involved.

3.4 Invasive Alien Species

Few natural or modified ecosystems are free from species that do not occur there naturally, and that have arrived or been introduced from other parts of the world. Many of the worst agricultural and other pest insects are alien to the areas where they have their greatest impacts, and may demand complex and expensive control to limit economic damage and loss. Insects are the most pervasive invasive species, but estimates of the costs of their impacts are difficult to make; Bradshaw et al. (2016) extrapolated from reported losses to state that invasive insects cost a minimum of US$70.0 billion a year globally, with associated health costs greater than US$ 6.9 billion/year. Many such alien animals and plants are 'naturalised', abundant, and widespread, are regarded as normal components of diversity in their novel environments, and continue to spread into natural ecosystems. Their interactions with native insects and others sometimes constitute threats. However, reflecting a very widespread scenario, the impacts of alien species on native invertebrates are overshadowed by concerns for the fates of native vertebrates, and are rarely investigated specifically. As examples, foxes and feral cats are amongst the most destructive predators of small mammals and birds in Australia, but their impacts on invertebrates are far less heralded. However, from 87 Australian feral cat diet studies, Woolley et al. (2020) found invertebrates in 39% of scat samples, with Orthoptera the most frequently represented order. Across the continent, mean annual consumption of invertebrates by feral cats was estimated at 1086 million individuals, with the highest proportion in largely natural environments. The impact of this is difficult to assess, but Woolley et al. suggested that cats could pose a threat to some large-bodied localised invertebrate species. Impacts of some other invasive vertebrate species on native insects are equally unquantified. The South American Cane toad (*Rhinella marina*), currently expanding its range across northern Australia, undoubtedly devours vast numbers of insects as part of its broad diet. However, almost all attention to the toad has focussed on its toxicity as prey of native reptiles, notably varanid lizards and other vertebrates susceptible to poisoning (Feit et al. 2018), and that have declined since toads invaded.

Two species of introduced trout are implicated widely in declines and losses of native fish, but may also threaten aquatic invertebrates. The spread of predatory Rainbow trout (*Oncorhynchus mykiss*) and Brown trout (*Salmo trutta*)—both species introduced and augmented from hatchery stocks for recreational fishing—in the Kiewa River, Victoria, is listed amongst the threats to the Alpine stonefly (*Thaumatoperla alpina*, p. 148), and may have caused losses amongst invertebrates at lower elevations. Elsewhere, alien fish have been implicated as a threat to dragonflies by larval predation (they, for example feed on some dragonflies on islands such as the Hawaiian archipelago, but perhaps not always as much as suggested [Englund and Polhemus 2001]).

Impacts of alien vertebrates are sometimes more conspicuous on isolated island habitats. The Black rat (*Rattus rattus*) almost certainly caused the demise of the Lord Howe Island stick insect (*Dryococelus australis*, p. 125), leaving only a tiny residual population on the rat-free remote Ball's Pyramid. Eradication of rats is necessary before *Dryococelus* can be reintroduced to Lord Howe. More widely, eradication or extensive management of alien vertebrates primarily for conservation of vulnerable native vertebrates may have many benefits to insects in the same areas. Suppression of introduced mammal predators in New Zealand is a cornerstone of conservation programmes for native birds as well as threatened Orthoptera (Weta, New Zealand's major flagship group of insects, p. 114), with mammal eradication from islands to leave them as predator-free 'havens' (p. 128) to which translocations can be made, as a central theme in weta conservation.

Isolated oceanic island faunas are particularly vulnerable to impacts of alien species. The spread and impacts of introduced alien insects in Hawaii is a well-publicised example (Howarth 1991), but parallels arise in many places. Some aspects of alien insect species impacts arouse serious concerns: they may be amongst the most serious pests present and demand substantial counter-measures. They are in general extremely difficult to eradicate, and some campaigns to do so have developed over many years of effort.

Social insects are amongst the paramount concerns, with their massive ecological threats to the receiving communities compounded by economic and health impacts on human welfare. Several ant species are listed amongst the world's most damaging alien invaders, because (as with some social wasps and bees) they spread rapidly, occur in large numbers, breed rapidly, and are efficient competitors and/or predators; some also sting, with severe pain and occasional more serious outcomes, associated clearly with entomophobia (p. 84). In assessing the most environmentally harmful alien insects in Australia, McGeoch et al. (2020) found 16 of 17 to be social insects, and ants predominant—nine species of ants, four of bees, and three of wasps (the other was the African black beetle [*Heteronychus arator*, Scarabaeidae], a major pest in turfgrass of lawns, golf courses, and other grassed areas, and a polyphagous feeder on many crop plants). Some ants, in particular, can substantially change the functioning and composition of ecosystems, attributed largely to the vast abundance they can achieve. The variety and impacts of social insects are evident from several cases summarised briefly below and selected to exemplify the major groups of social insects that have caused concern in Australia or New Zealand.

1. The honeybee, *Apis mellifera*, was introduced to Australia early in the nineteenth century (1822) and is an important pollinator of many crops and ornamental plants, as well as a source of honey and wax. As well as domesticated hive populations that found the apiculture industry, honeybees have become feral over much of temperate Australia, with suggestions that they may outcompete some of the numerous native bees when nectar sources are sparse. Clearing of forests over large areas has reduced nectar availability from eucalypts, although forest persists in protected areas such as national parks. Much commercial apiculture is based on 'migratory beekeeping', the seasonal movement of hives to track nectar sources or commercial pollination demands, and some scientists have

recommended that hives should not be placed in or near protected areas for fear of adverse interactions with native pollinators.

'Managed bees', both native and alien, are critical for crop pollination in many parts of the world, and fear of their adverse impacts is widespread. These fall into three main categories: (1) competition with native bees for nesting sites and floral resources; (2) indirect changes through changes in plant communities—including declines of native flora and spread of alien species; and (3) transmission of pathogens. A global review of the impacts of honeybees and bumblebees, as the most intensively managed taxa, by Mallinger et al. (2017) concluded that further research was needed to understand the outcomes thoroughly, but impacts varied according to whether managed bees were in their native or non-native range, with the former having lesser competitive effects but potentially greater impacts on wild bees through pathogen transmission. More than 100 other bee species are potentially manageable, and include a number of solitary taxa; many are important for particular crops (Graham et al. 2023).

2. Four species of bumblebees, *Bombus* spp., were introduced to New Zealand from Europe in the 1880s as pollinators, especially for Red clover. New Zealand and Australia have no native bumblebees, but *B. terrestris* was found in Tasmania in 1992. Apparently, two queen bees from New Zealand reached Battery Point, Hobart, and were presumably imported on boats. It spread rapidly and the bumblebee now occurs throughout the island state and is often the most abundant and conspicuous bee present, even in remote native floral associations. Only a few years later, reports that *Bombus terrestris* could displace native megachilid bees (*Chalicodoma* spp.) implied wider effects on native pollinator efficiency and wellbeing in Tasmania (Hingston and McQuillan 1999). Its adverse impacts are thought to be considerable, through (i) its habit of buzz pollination, a contrast to all native pollinators and (ii) chewing the flower corollas to obtain nectar, so damaging the individual flowers. However, increased 'buzz pollination' is a strongly advocated need in glasshouse production of crops such as tomatoes, so that suspected high ecological damage from the bee is accompanied by strong calls for it to be imported to the Australian mainland (especially New South Wales) for glasshouse use. Fears of escape have so far prevented this, and 'the release and spread of the Large earth bumblebee, *Bombus terrestris*' is listed formally as a 'potentially threatening process' in Victoria. A wider Australian Senate Inquiry into use of bumblebees in Tasmania for commercial pollination later lapsed but the scope of the proposed inquiry indicated the wide range of issues considered relevant (Table 3.3).

3. The alien European wasp, *Vespula germanica* and Common wasp, *Vespula vulgaris* are both predators of small arthropods. However, adult wasps are omnivorous, with opportunistic diet including honeydew, nectar, a wide range of (mostly) invertebrate prey and carrion. Unlike the annual development in their native European range, where colonies remain small and do not overwinter, development in the warmer climates of New Zealand and Australia is not as

Table 3.3 The terms of reference for the proposed inquiry into 'The risks and opportunities associated with the use of the bumblebee population in Tasmania for commercial pollination purposes' (Australian Senate 2016)

The existing distribution and population density of exotic bumblebees
Productivity and economic benefits of the commercial use of bumblebees for agricultural producers.
The potential environmental impacts associated with the commercial use of bumblebees, including whether their use is likely to:
• Impact the conservation status of a species or ecological community.
• Impact biodiversity.
• Cause unintended ecological impacts.
• Contribute to a wider distribution of bumblebees.
The implications for Australia's biosecurity regime of any approval to use bumblebees in Tasmania for commercial purposes.
The potential economic outcomes.
The effectiveness of alternative pollination options.
Any other related matters.

seasonally constrained and very large perennial nests can be produced, with up to thousands of queens; indeed, reports exist of nests being 4 m in length and containing millions of wasps (Spradbery 1988). The highest known wasp densities are reported from New Zealand's beech forest, where densities of up to 170 individuals/m^2 of tree trunk feed on honeydew produced by an endemic scale insect (*Ultracoelostoma assimile*), with one estimate reporting a peak of 3761 g/ha of *Vespula* biomass (see review by Lester and Beggs 2019), sometimes far exceeding that of birds and introduced rodents and mustelids. The wasps deplete arthropod populations and affect food supply for birds, as well as killing nestling chicks, leading to long-term declines of some bird species. Most moth caterpillars placed experimentally in beech forests with high wasp populations did not survive beyond 4 h; such protein resources are needed throughout the wasp's life cycle. Lester and Beggs also summarised a range of sublethal effects, through resource depletion and aggressive behaviour, collectively leading to widespread impacts on community structure and ecosystem functions. Loss of pollinators (through exploitation and competition), food, and other key community inhabitants are major concerns from invasive wasps.

V. germanica, now a widespread invasive species in Tasmania, has been implicated as a threat to the native Ptunarra brown butterfly (*Oreixenica ptunarra*) as a predator of early stages (Woinarski et al. 2023). The wasp is also widespread on the southeast mainland.

4. The Yellow crazy ant, *Anoplolepis gracilipes,* is one of the world's most invasive species, and gets its name from its pale colour and its generally frenetic activity. It has spread widely and 'has destroyed local ecosystems in numerous countries' (Jung et al. 2017), with further spread anticipated through human activities; it is a significant threat to biodiversity. The ant can form vast 'supercolonies' that can

exceed 10,000 ha in extent (Hoffmann 2014), containing thousands of queens, and is an aggressive polyphagous predator, attacking and killing such large animals such as piglets, dogs, and chickens in domestic environments, as well as nestling birds, reptiles, and many invertebrates elsewhere. It is perhaps best-known in the Australian region for mass killings of indigenous migratory Red crabs (*Geocarcoidea natalis*), a keystone species on Christmas Island; it also blinds crabs through spraying acid onto their eyes, but one estimate suggests that more than 20 million crabs have been killed by the ant. Ants also take carbohydrates from plant nectaries and honeydew secreted by scale insects and aphids on vegetation. *A. gracilipes* is amongst a number of so-called 'tramp ants' that have become widespread and are of considerable concern over their varied but significant impacts—concerns for Australian biodiversity are widespread (Commonwealth of Australia 2006), and threats recognised under the Environment Protection and Biodiversity Conservation Act's listing of the ant on Christmas Island as a Key Threatening Process in 2005.

5. The Red imported fire ant (RIFA), *Solenopsis invicta,* is an equally significant invader, of concern mainly for its impacts on human health in addition to being a serious agricultural and environmental pest. Its discovery in Australia (in two sites in Brisbane in February 2001) led to a detailed surveillance campaign (from which it was inferred to have then been present for 3–5 years but already occupied more than 40,000 ha: Vanderwoude et al. 2003) and an eradication plan. Climate modelling implied that RIFA could potentially colonise much of Australia other than deserts and high-elevation areas. Its significance was recognised by the listing of 'The reduction in the biodiversity of Australian native fauna and flora due to the red imported fire ant …' under the Environment Protection and Biodiversity Conservation Act 1999 in 2003, with a variety of native birds and others listed as potentially vulnerable to the ant's attack.

Parallel concerns over impacts of social Hymenoptera occur elsewhere. Invasions of European bumblebees, following introductions to Chile in the 1980s–1990s have been implicated in the severe decline of the Giant Patagonian bumblebee (*Bombus dahlbomi*), the world's southernmost bumblebee and the sole native forest bumblebee in temperate South America, where it was an important pollinator (Morales et al. 2013, 2022).

International trade in honeybees and bumblebees for pollination services is frequent, with the consequent spread of bee viruses and other pathogens that can devastate hives, and threaten the apiary industry. Many of these organisms are now widespread. Until recently, Australia was the only major beekeeping region in the world without the mite *Varroa destructor*, an ectoparasite of honeybees, but it was discovered near Newcastle, New South Wales, in 2022, stimulating a rapid quarantine and surveillance response, with destruction of infested hives and sentinel hives deployed for monitoring. In August 2023 *Varroa* spread to occur in the Riverina district of western New South Wales, with fears that it is now established in Australia and will be difficult to eradicate.

The variety of invasive alien insect species and the contexts of their establishment extend far beyond social species. The non-target impacts of classical biological control agents are a frequently expressed concern, and can be problematical for conservation if the agents spread or are inadvertently moved into natural areas. Indeed, some early commentators queried the compatibility of this practice with successful insect conservation (Samways 1988). In the past, many such agents were introduced from the pests' areas of origin in the expectation of controlling alien plant weeds (as herbivores) or arthropod pests (as predators or parasitoids: natural enemies) without rigorous testing of their specificity, and some subsequently changed or expanded habits to exploit native plants, prey or hosts in their release areas—in some cases causing concerns for the fate of their victims, or contrary outcomes from their activities in different places. The South American pyralid moth *Cactoblastis cactorum* spectacularly diminished invasive prickly pear cactus (*Opuntia* spp.) in Australia, but later (and in a very different ecological context) proved a serious threat to rare cacti in North America, where conservation measures for the cacti included suppression of the moth.

Specificity of any introduced agent is often not fully known, but pre-introduction investigation has now become routine and obligatory. The tachinid fly *Compsilura concinnata* was introduced to North America from Europe in 1906 to control Gypsy moth (*Lymantria dispar*, an important forest defoliator) but, because of the limited seasonal availability of this host, needed to use native Lepidoptera to sustain its populations over much of the year. Elkinton et al. (2006) noted that it had been recorded from more than 180 native species, and that it was implicated also in the near extinction of another invasive moth pest, Browntail moth (*Euproctis chrysorhoea*, Erebidae: Lymantriinae). Major conservation concerns arose from the fly attacking native silkmoths (Saturniidae), with substantial declines of some species.

Extensive testing of any proposed biological control agent before it is approved for release is designed to assess its specificity and assure that it is 'safe' for release. The process eliminates many candidates, and increases the chances that any introduced agents will indeed have only minimal non-target effects. However thorough the screening undertaken to assure 'safety' before they are released, the possibilities of classical biological control agents later adopting native, non-target species cannot be wholly discounted. Any such biocontrol agents are likely to become permanent additions to the receiving areas, and various attempts to eradicate them rarely succeed when they spread into non-natural environments from their intended sphere of impact.

Introduced commercial or ornamental plants may become resources for native insects—some with unexpected consequences. The garden ornamental South American Dutchman's pipe vine (*Aristolochia elegans*) spread widely into forests in southeastern Queensland and was an attractive oviposition substrate for females of the Richmond birdwing butterfly (*Ornithoptera richmondia*). However, young larvae of the butterfly were poisoned by eating the vine foliage and died. Through this ecological trap, the vine was implicated in the severe decline of this spectacular butterfly, together with habitat decline by clearing of forest, and a key component of a long-term conservation campaign was to remove this alien vine from forest and undertake extensive planting of the native food vine (*Pararistolochia praevenosa*) from nursery-grown stock (Sands and New 2013), so that female butterflies had easier access to a suitable larval food plant.

Likewise, introduced weeds may be exploited by native insects, and their role leads to mixed appreciation of their presence. Chilean needle grass (*Nassella neesiana*) has invaded native grasslands in southeast Australia and is a declared noxious weed—a formal status that obliges eradication wherever possible. However, *Nassella* may also be an important larval food plant for the Golden sun-moth (*Synemon plana*, p. 157) in areas where native grasses are sparse, so its presence may then be a conservation benefit and lead to contrasting management priorities.

Effects of alien plants may be much more complex than those from food alone. Waterside invasive alien trees (mainly Australian species of *Acacia*) threaten most Red-Listed Odonata species in South Africa, largely by shading causing loss of needed sunny environments, but also with subsidiary effects such as bank erosion from loss of native vegetation. Once those trees are removed, most odonate species recover rapidly (Samways and Sharratt 2010). As another example, dense floating mats of invasive aquatic weeds such as Water hyacinth (*Eichhornia crassipes*) can eliminate native species and reduce heterogeneity of lake and pond environments, with consequent losses of specialist aquatic insects.

Some aspects of alien insects and other introductions are extremely difficult to monitor or regulate. The harmful 'tramp ants' noted above are a salutary warning of less obvious arrivals, with the pet trade in insects an increasing concern. The trade in ants as pets has increased greatly since the early 2000s, facilitated by ease of international transport through the postal service (so that small packages including a queen, several workers and some brood in a small vial or package are moved around cheaply). Whilst some invasive ants have been spread through the global commodities trade or accident, the pet trade includes many species regarded as invasive. Gippet and Bertelsmeier (2021) noted that at least 520 ant species (representing 95 genera) were available for sale online between 2002 and 2017. These included 57 invasive species, including 13 of the 19 worst global ant invaders, all likely to have high economic or ecological impacts should they establish outside their natural range, through escape or discard. Many are not identified precisely or are named erroneously (New 2005), and Wang et al. (2023) also noted the high invasion risk from the ant trade. Their survey in China reported sales of 209 species in a 6-month period, with many of the most sought-after species having high invasion potential. Indeed, from Gippet and Bertelsmeier's (2021, see also Gippet et al. 2022) analysis, the ant trade favours invasive species, these being (1) species with multiple queens in a nest; (2) large in size; (3) habitat generalists; (4) nesting in a variety of habitats; and (5) with large range sizes—these attributes summing to constitute effective and adaptable generalist species. One contrast to this is Lester's (2005) finding from a survey of New Zealand's alien ants that small size might strongly favour establishment.

The pet ant market is little controlled, and several authors have expressed deep concerns over the ease with which serious pest ant species—including some already the targets of expensive and complex eradication attempts—can be moved across international borders or taken to places outside their natural range (Lassaline et al. 2023). Some controls do indeed exist. Ant dealers in eastern Australia, for example cannot export to the United States or to Western Australia or Tasmania within Australia. Nevertheless, it appears easy to move native ants within the country to parts of Australia where they do not occur naturally, and for the identity of such transplants to be uncertain. Several highly venomous species are also advertised for sale, such as some bullants from Australia; queens of various *Myrmecia* species are priced in the range of Au$200–300 each in recent website listings (in May 2023). They include *Myrmecia pilosula*, the Jumping jack ant, that has caused human deaths (through anaphylactic reaction) from its sting—in some cases with advertisements containing little warning of its potential impact. *Myrmecia* are attractive to potential purchasers as amongst the largest and most spectacular looking ants advertised, and are restricted to Australia and New Caledonia, but their establishment elsewhere could be a concern: an early establishment of *Myrmecia brevinoda* in New Zealand was eradicated, with follow-up surveys including advertising an award (not taken up) of $100 for specimens found (Lester and Keall 2005).

3.5 Overexploitation

Overexploitation refers to the excessive take of organisms to a level that the population cannot replace and leads to net loss, potentially increasing the risk of extinction. Many common insects are captured in large numbers, for tourist ornaments and similar purposes, and this practice generally raises little concern. Nevertheless, the collection of insects that are harvested for food, medicine, pets, fishing bait, and other commercial purposes is difficult to monitor and regulate. However, commercial exploitation of both spiders and insects can represent a serious threat to rare and 'desirable' species (Losey et al. 2022) and the more frequent conservation concerns arise from collectors seeking specimens of demonstrably rare insects for their personal collections or for sale. Formal listing of insects for protection generally includes some form of 'prohibition of take', whereby collecting of any specimens becomes illegal unless under a specific permit. Concerns arose initially from collecting of rare Lepidoptera, and that this might accelerate their path to extinction by providing another threat to already small and vulnerable populations. Thus, the extinction of the Large copper butterfly (*Lycaena dispar dispar*) in Britain, attributed mainly to drainage of its fenland habitat, was perhaps accelerated by intensive collecting of this rare butterfly as it became scarcer (and more desirable!) in the years before loss. Accusations of losses from overcollecting are easy to make, but very difficult to prove. Several historical cases in the northern hemisphere have been cited widely. One such case is of the New Forest Burnet moth (*Zygaena viciae*) that occurred formerly in a few small sites in southern, England from which it was thought to have been lost from overcollecting pressure, becoming extinct in 1927. However, a population was discovered in Scotland in 1963, where it has been nurtured since; the precise locality has not been publicised to reduce risk of exploitation (Young and Barbour 2004, and representing a practice that elsewhere sometimes continues to cause debate), and no other populations have been found despite extensive searching. The excitement of rediscovering a presumed extinct insect may thus be tempered by need not to publicise that discovery for fear of vulnerability; the species involved are inevitably rare, increasing commercial desirability and price. Such rediscoveries of presumed extinct species of commercially desirable groups can stimulate or renew interest in exploitation, as reported for an endemic Chilean stag beetle (*Sclerostomulus nitidus*) rediscovered in 2012. Sampling by counting beetles in dead logs over 2013–2018 suggested a 93% decrease in abundance over that period (Crespin and Barahona-Segovia 2020), who also witnessed collecting and noted online trade for the beetle, despite its listing in Chile as 'Critically endangered'. Commercial plundering of rare butterflies and beetles can be increased by the notoriety of conservation importance, and can be very difficult to monitor or regulate. This 'high value - low volume' commercial trade is that of greatest concern in conservation.

Collecting for the pet trade has been implicated widely in declines of stag beetles (Lucanidae), but a variety of other insects are also sought. For ants, the primary concern is for invasiveness rather than exploitation.

For most insects, 'collector pressure' is far smaller and overcollecting is not a major general concern. However, listing and prohibition create desirability, and 'rarity' increases commercial value, leading to black market operations of so-called protected species, most notably butterflies and beetles. Greater formal monitoring and plans to mitigate online exploitation are needed (Losey et al. 2022). Some cases may transcend the more common trading picture of 'commercial extinction', whereby costs of obtaining the rarest species exceed the reward from selling them. Particular birdwing butterflies and stag beetles are the most sought insects in trade, and enormous prices are asked for the rarest species; all birdwings are listed on CITES because, whilst the general appearance of these butterflies is distinctive, many of the species are more difficult to distinguish—so that listing 'all birdwings' removes the need for non-entomologist customs officials to recognise individual taxa. Unusually amongst insects, stag beetles have two sequentially different markets, as pets and as pinned specimens. Many stag beetle species are quite variable, and the largest specimens with longest mandibles are the most desired and expensive. Perhaps particularly in Japan, concerns have arisen over expansion of the pet beetle trade, and the variety and abundance of rare stag beetles imported to satisfy this—with subsidiary concerns over releases or escapes leading to hybridisation with native species. Expansion of the 'insect pet' trade is difficult to monitor, as are identities of some species advertised. The Australian bull ant known as '*Myrmecia* sp. 17' is listed as threatened in Victoria, but 'small colonies' (queen and several workers) have been advertised on a commercial website (in November 2023: price $250).

The principle of prohibiting collecting (broadly 'take') of putatively rare, poorly known, or 'listed' insects arouses much debate. For many insects, the grounds for listing are flimsy, but prohibition is 'automatic' under well-intentioned legislation, with the implication that collecting constitutes a threat. Critics have pointed out that the step is often ill-founded, and alienates the interests of the very people who can contribute most to advancing knowledge of the species and formulating its conservation needs—and those same people may be those who initially nominated the species for protection out of genuine concern for its fate. That pool of knowledge, experience, and goodwill is a vital need in insect conservation and, indeed, butterfly conservation (the foundation of wider insect conservation interest) had its genesis in the concerns of hobbyists as species became scarcer. And, whilst 'butterfly collecting' has become stigmatised as socially unacceptable in parts of the northern hemisphere, much further collecting is needed over much of the rest of the world to characterise the poorly known faunas. Attempts to sustain hobbyist interest by imposing 'bag limits' rather than total prohibition may also be problematical: one responsible collector taking a single pair of a rare butterfly may be of little consequence, but many equally responsible collectors each capturing a pair may accumulate the overall take to a harmful level.

The historical accumulation of distribution records of rare Lepidoptera (in particular) in collector-rich countries such as England is a foundation for much of the recent concern over vulnerability: however, some records are misleading, and may never be elucidated. Allan (1943) recounted how some collectors would deliberately mislabel specimens of the rarest (and most valuable) species to avoid 'their' localities being disclosed to rival hobbyists! The practice can have serious practical consequences as today's scientists may waste resources exploring such places, assuming the records are genuine and that the species involved may have become extirpated.

The rise of photographic recording of insects on platforms such as 'iNaturalist' (p. 16) is welcome, but for many groups identification cannot be confirmed reliably without access to the specimen. In short, regulation of insect collecting activities is often contentious—notwithstanding widespread acceptance that overexploitation should not occur.

Many insects are harvested for human food (Chap. 6), some with a long tradition of use as a protein source, and the global value of wild insects in human diet continues to grow. Australian First Nations people have for long harvested honey ants and 'witchetty grubs' (broadly, a name applied to the large woodboring larvae of some cossid moths and beetles) for their own use and sale, and recent increase in tourism in inland Australia, accompanied by interest in sampling 'bush tucker', is increasing demand for these insects. Fears of local depletion of these long-lived insect larvae (that may take several years to develop) have been expressed. In southern Africa, harvesting of highly nutritious mopane worms (caterpillars of an Emperor moth, *Gonimbrasia belina*, Saturniidae) has been associated with local extirpations. Unlike the cryptic larvae of witchetty grubs, mopane worms feed exposed in foliage, and are much more seasonal in incidence. Harvesting is most intense during November–December, with a smaller second period in April–May. However, as well as being seasonal, production is also 'erratic and unreliable' (Bara et al. 2021), because of parasitisation and rainfall vagaries. Bara et al. noted that absence of any monitoring or regulatory framework that would protect *Gonimbrasia* from overexploitation is also a major concern. The nutritional importance of mopane worms as a major protein source and that their harvest provides livelihoods for many people, has led to suggestions for facilitating year-round commercial production. Many aspects of entomophagy impacts on natural insect populations are unknown—but, for example the extent of harvesting and sale of edible ants and their brood is substantial (Rastogi 2011), and has led to some species being regarded as threatened. Larvae and pupae of social wasps are important human food items in at least 19 countries, and wasp-nest gathering for sale is a livelihood and income source for local farmers (Brock et al. 2021). Bulk harvesting of aquatic fly larvae and/or puparia, notably of blackflies (Simuliidae), midges (Chironomidae), brineflies (Ephydridae), and mosquitos (Culicidae) have all contributed to traditional entomophagy in various parts of the world.

The use of insects in traditional medicine is also widespread (Chap. 6), often at low intensity, but with occasional fears of overharvesting. Together with climate change (below), overharvesting of caterpillar fungus (*Ophiocordyceps sinensis*, sometimes referred to as one of the world's most valuable biological commodities) is causing major declines over the Tibetan Plateau of China, and in the Himalayan areas of Bhutan, Nepal, and India (Hopping et al. 2018). Declines of the fungus have been reported across its range, so that harvesting has become more laborious. The fungus parasitises caterpillars of more than 50 species of local hepialid moths (mostly *Thitarodes* spp.) restricted to high-elevation grassland areas, and their harvest (by individual collecting) provides 'a crucial source of income for hundreds of thousands of collectors' (Hopping et al. 2018). The subterranean caterpillars feed on grass roots and take several years to develop. It is feared that continued demand for this high-value commodity will not be sustainable, in conjunction with environmental degradation from climate changes and grazing regimes. Over much of the region, caterpillar fungus sales for medicinal use are a substantial component of gross domestic product, and a valuable export: Hopping et al. quoted sale prices of more than US$ 140,000/Kg in Beijing in 2017, and Wang et al. (2022) estimated the Asian market value of US$ 5–11 billion. International trade borders and permissions are confusing so that the trade is enormously difficult to monitor or regulate, and uncertainties have led to listing of *Ophiocordyceps* as 'Vulnerable' on the IUCN Red List (Yang 2020).

Use of insects as bait for recreational fishing can lead to local concerns over the use of dragonfly larvae and, especially in Australia, larvae of some hepialid moths dug up from around tree roots—as with witchetty grubs, these may demand prices of several dollars each for local sales to recreational fishers. Large dragonfly larvae ('mudeyes') are also a popular bait.

Although not usually included in conservation concerns for the source populations, the mass collections of biological control agents such as some ladybirds for release elsewhere, and releases of butterflies as aesthetic and ceremonial attractions at weddings and other festivals can suggest unwanted impacts in the release areas. The numbers released can be very large—Losey et al. (2022) quoted annual totals of 3000 000 000 individuals of the Convergent ladybird (*Hippodamia convergens*) and 9,900,000 individual Painted lady butterflies (*Vanessa cardui*) sold for pest management and weddings, respectively, in the United States. Losey et al.

suggested that the massive total of 3 billion ladybirds collected each year from the Central Valley of the Sierra Nevada a year might not be sustainable in the longer term—reasoning based on the following presumptions from various sources: (1) every acre of alfalfa in the Valley can produce around 50,000 ladybirds a year; (2) the total area of alfalfa in the Valley is around 500,000 acres, giving an extrapolated total of around 25 billion ladybirds; (3) a 'reasonable estimate' that about half these will survive the winter, leaving around 13 billion; and (4) these estimates did not include losses from migration, pesticides, fire, and other factors, all of which could erode pest management by *H. convergens* in the Valley and endorse needs to retain substantial populations.

3.6 Climate Change

As small-bodied ectothermic beings, many insects depend on very restricted climate or microclimate regimes, and any changes to these affect their survival, rate, and timing of development, behaviour, physiology, and interactions with other taxa. They are considered susceptible to changing temperature and moisture regimes, and changes to these can have severe impacts on insect wellbeing. Climate change is thus a universal superimposition on all other threats, and its impacts are difficult to appraise or predict. It also differs from most other threats in that individual people and groups can do nothing directly to thwart its influences, other than lobbying for governmental policy changes, and must focus on aspects of habitat protection and connectivity.

Climate warming is considered widely to be the most serious stress on the natural environment flowing from human activities. It is likely to affect most ecosystem structures and services influenced by invertebrates as their environments and local assemblage compositions change, and lead to changes in the richness and abundance of local biodiversity in many subtle ways. A study of the effect of climate warming on ants, for example suggested that they might move more soil and build deeper nests to avoid increasing temperatures, but warming may also influence their effects on soil ecosystem processes (Del Toro et al. 2015). Cases of insect declines attributed to climate changes span many different environments. Forest floor beetles declined in abundance and diversity over 45 years in the Hubbard Brook Experiment Forest, New Hampshire (USA). Nineteen families disappeared, most notably predaceous beetles (Harris et al. 2019) amongst which Staphylinidae: Pselaphinae—the most abundant group in 1973–1977 samples—were not found in 2015–2017. Changes were attributed to reduced depth and duration of snow cover over this period, leading to soil freezing to greater depths—so that winter warming led to harsher conditions for the forest floor fauna.

Harvey et al. (2022) distinguished two major categories of climate change impacts on insects as (1) gradual long-term events (warming, seasonality) leading to range shifts, seasonal changes and affecting species interactions; and (2) extreme events (extremes of temperature and rainfall, each with parallel direct and sequential impacts such as fire, drought, desiccation, and flood), each of which can have massive and poorly understood impacts.

Common outcomes from warming over recent decades have included distributional and seasonal changes—with many insects reported to be colonising higher latitudes or higher elevations, some then unable to persist in the opposite extremes of their former range as conditions there change or resources become unavailable. The uneven combination of shifting between the 'leading edge' of expansion and 'trailing edge' of abandonment may lead to overall range losses. Thus, as discussed by Morales et al. (2022), many European and North American bumblebees (*Bombus* spp.) have disappeared from southernmost parts of their ranges, with these not compensated by equivalent colonisation of novel northern or upper areas. Insect species' distributions are labile, and many have changed in response to warming, but any single species' change has consequences for others and for the community as a whole. Many specialist insects may be vulnerable to being displaced by more tolerant taxa invading 'from below'. Warming of small streams is projected as a major threat to some narrowly endemic stoneflies in Australia's alpine regions and that are adapted to life in these cold waters (p. 148) and are already rare. More generally, many cold-adapted insects in southern Australia may prove vulnerable to climate warming. In North America, species richness of forest ants was predicted to decrease in southern regions and increase in northern regions, from a combination of thermal stress (south) and increased habitat suitability (north) (Del Toro et al. 2015), accompanied by changes in the functional categories (p. 105) of ants present. *Ophiocordyceps sinensis* (above) on the Tibetan plateau apparently needs winter temperatures well below 0 °C and to avoid permanently frozen soil, with warming also affecting the grassland habitat: Hopping et al. (2018) noted that simultaneous shifts in distribution of the *Thitarodes* caterpillars, the fungus, and the necessary vegetation and soil conditions on which they depend will not necessarily occur. The fate of the local *Thitarodes* species is also uncertain.

Northward movements, especially of Lepidoptera and Odonata in Europe (Platts et al. 2019), are far better documented than southward movements of insects in Australia (for example) but there is little reason to suppose that such parallels do not occur. They are difficult to predict or model—not only from lack of information but because a species' capability to move successfully depends on many aspects of its biology, rather than just its ability to disperse. Upward movements to invade montane or alpine zones pose threats to the restricted high-elevation insect faunas encountered, as novel interactions occur—and, in many cases, insects already living at high elevations and also influenced by climate change cannot move to avoid them—they have 'nowhere to go' as their habitats decline and become increasingly isolated. Some alpine Lepidoptera in Australia, for example depend on (and are confined to) high-elevation natural and semi-natural open grassland areas, including frost hollows, which are likely to be progressively lost to encroaching woody vegetation. Alpine aquatic insects that depend on cold streams may be susceptible to warming as snow loss increases, as Giersch et al. (2017) discussed for some North American stoneflies with very narrow distributions in alpine streams.

As Bush et al. (2014) noted, the limited elevational range of Australia ensures that many insects will need to undertake latitudinal range changes as their 'climate envelope' shifts, with dispersal ability an important parameter of success. In modelling the distribution changes of Australian Odonata for several future climate scenarios, Bush et al. assessed three components relevant to the vulnerability of the 270 dragonfly taxa studied. These components (Fig. 3.1) are (1) exposure: the extent to which a species' currently occupied physical environment will change; (2) sensitivity: the extent to which suitable habitat is lost; and (3) dispersal pressure: the reliance on dispersal to avoid further negative impacts. Species susceptible to all three components were deemed 'vulnerable', but the three other possible impact combinations are noted in the figure. The trends found from the modelling across several change scenarios included (1) 55–69% of species predicted to undergo overall decline in habitat extent by 2085; (2) environmental conditions shifted beyond those experienced by 50–95% of species in their current suitable range; and (3) most species were predicted to shift towards higher altitudes. A third of modelled species were considered to become highly vulnerable or vulnerable by 2085 under some regimes tested. Bush et al. (2014) concluded that climate change presents 'clear danger to Odonata and other freshwater species'. Some trends, likely to apply far beyond Odonata, are difficult to predict—thus, Bass Strait may be an effective barrier for natural dispersal of many insects that could thrive in Tasmania, so that the southernmost mainland is 'as far as they can go'.

Rates of insect movement in range changes vary considerably across species and are difficult to predict. Again as exemplified by Odonata, some species are strong migrants and can cover considerable distances rapidly: these are unlikely to

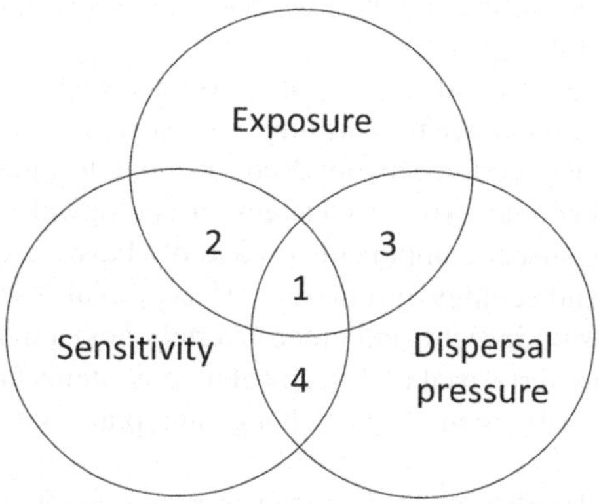

Fig. 3.1 The categories of vulnerability to climate change, based on the three components of exposure, sensitivity, and dispersal pressure and used for Australian dragonflies. Numbers suggest different levels of risk, as: (1) Highly vulnerable; at greatest risk; specific research or targeted action generally needed. (2) Vulnerable; at high risk; significant within-region decline; identify refugia and manage other threats. (3) Potential to persist; may not be at risk; monitor populations and dispersal trends. (4) Theoretical risk: highly unlikely; species habitat suitability both declines and shifts prior to high environmental exposure (after Bush et al. 2014, used under CC-BY licence conditions)

succumb to climatically changing environments. Many others are regarded as poor dispersers, and Bush et al. (2013) noted that Australian dragonfly species distribution 'appears highly sensitive to climatic factors', so that Odonata might be useful as indicators for climate change responses. Distances moved by non-migratory species are reflected in the northward movements reported in Britain, where 37 non-migratory dragonfly species expanded northward by an average of 6.8 Km/year (Hickling et al. 2005).

Many such range changes also necessitate abandoning the trailing edge of the former range as conditions there become unsuitable—either because of warming or because of reflecting different effects on an insect and its key resources or changed interactions brought about by other species invading those areas. Climate may affect seasonal development and the availability of a food plant and its insect consumer at different rates, so that long-founded and finely honed obligatory interactions become 'decoupled' by changed phenology or tolerances. Earlier seasonal appearance of some butterflies and others has been attributed to climate warming. Increased periods of suitable temperatures for development (such as 'longer summers') may enable an extra insect generation to occur—but may also act as an ecological trap if individuals from a partial extra generation are unable to reach a 'safe' growth or diapause stage and perish by the time conditions deteriorate. Changing diapause regimes may also complicate insect development and synchrony with foods, as developmental cues from climate signals change. Changes to overwintering strategies, for example may be both complex and their consequences far-reaching and severe. Resuming development before key resources are available must inevitably be a 'developmental trap' leading to mortality through mismatch with needs. Thermal mismatches between insect and plant, or insect and its natural enemies (parasitoids, predators, pathogens) have the potential to disrupt complex food webs and their embedding interactions.

Essentially, insect communities are undergoing change, with these most pronounced at higher latitudes where warming is most rapid. Unless populations and species adapt to the new conditions imposed, or move to inhabit new temperature regimes, they may become extinct. Changes in ecological communities flowing from changes in their insect components potentially have severe and wide-ranging impacts on stability and ecological functions (Harvey et al. 2022). The combination of short-term impacts on insects (and other invertebrates) and the consequent long-term impacts on many terrestrial and freshwater ecosystems emphasise the need to understand the ramifications of climate change and plan to mitigate impacts as far as practicable.

In large part linked with climate change, increased frequency of 'extreme weather events'—floods and droughts, the latter associated with increased wildfire intensity following periods of elevated temperatures—can dramatically affect the areas in which they occur, and cause massive invertebrate mortality and ecological change, beyond immediate impacts such as deaths from heat stress or drowning, physiological changes, and losses of key resources. Recent unprecedentedly intense wildfires in Australia and elsewhere have destroyed or threatened human life and property, with losses of fauna incalculable. An attempt to enumerate the temperate rainforest litter terrestrial invertebrates destroyed in the severe 2019–2020 fires (which consumed about 97,000 Km^2 of vegetation in southeast Australia; another estimate was of 7.2 million ha of land area [Silva et al. 2020]) implied that an overall 60 billion individuals perished (Gibb et al. 2023). A year after the fires, the highest-intensity burn sites supported only 26% of the invertebrate abundance of unburned sites, and areas subjected to these highest-severity fires supported 1.90 million fewer individuals per hectare. Although many taxa persisted, substantial declines occurred. The responses of insects to natural low-intensity fires vary greatly, depending on taxa and the heterogeneity (such as availability of refuges) of the sites affected, but high-intensity burns from 'megafires' override this capability and render all taxa vulnerable, across most terrestrial and freshwater environments. In general, bushfire impacts on Australian invertebrates are not well understood, so that their responses are very uncertain to predict. Thus, Silva et al. (2020) noted for the 2019 megafires that the combination of fire severity, high elevations of many fires—so, steep slopes—and increased run-off from rainfall washed sediment, ash, and debris with a range of chemicals was likely to have detrimental impacts on aquatic biota well beyond the fire zone itself. Likewise, Saunders et al. (2021) noted that distant impacts could include transport of ash into marine systems thousands of kilometres from the actual fires. They also noted mass 'wash-ups' of dead insects on beaches far downwind of the fires. Many of the conflicting reports of prescribed fires having positive or negative impacts on insects (see New 2014) may flow from inadequate attention to the influences of fire severity (Mason et al. 2023).

References

Allan PBM (1943) Talking of moths. Montgomery, Newtown

Alyokhin A, Nault B, Brown B (2020) Soil conservation practices for insect pest management in highly disturbed agroecosystems—a review. Entomol Exp Appl 168:7–27. https://doi.org/10.1111/eea.12863

Ancillotto L, Labadessa R (2023) Functional traits drive the fate of Orthoptera in urban areas. Insect Conserv Divers 2023:1–8. https://doi.org/10.1111/icad.12683

Bara GT, Sithole R, Macheka L (2021) The mopane worm (*Gonimbrasia belina* Westwood): a review of its biology, ecology and utilisation in Zimbabwe. J Insect Food Feed 8(8):823–836. https://doi.org/10.3920/JIFF2021.0177

Boyes DH, Evans DM, Fox R, Parsons MS, Pocock MJO (2021a) Is light pollution driving moth population declines? A review of causal mechanisms across the life cycle. Insect Conserv Divers 14:167–187. https://doi.org/10.1111/icad.12447

Boyes DH, Evans DM, Fox R, Parsons MS, Pocock MJ (2021b) Street lighting has detrimental impacts on local insect populations. Sci Adv 2021(7):eabi8322

Bradshaw CJA, Leroy B, Bellard C, Roiz D, Albert C (et al.) (2016) Massive yet grossly underestimated global costs of invasive insects. Nat Commun 7:12986. https://doi.org/10.1038/ncomms12986

Brock RE, Cini A, Sumner S (2021) Ecosystem services provided by aculeate wasps. Biol Rev 96:1645–1675. https://doi.org/10.1111/brv.12719

Bush A, Theischinger G, Nipperess D, Turak E, Hughes L (2013) Dragonflies: climate canaries for river management. Divers Distrib 19:86–97. https://doi.org/10.1111/ddi.12007

Bush A, Nipperess DA, Duursma DE, Theischinger G, Turak E, Hughes L (2014) Continental-scale assessment of risk to the Australian Odonata from climate change. PLoS One 9:e388958. https://doi.org/10.1371/journal.pone.0088958

Cartright J (2019) Ecological islands: conserving biodiversity hotspots in a changing climate. Front Ecol Environ 17:331–340. https://doi.org/10.1002/fee.2058

Classen-Rodriguez L, Tinghitella R, Fowler-Finn K (2021) Anthropogenic noise affects insect and arachnid behavior, thus changing interactions within and between species. Curr Opin Insect Sci 47:142–153. https://doi.org/10.1016/j.cois.2021.06.005

Commonwealth of Australia (2006) Background document for the threat abatement plan to reduce the impacts of tramp ants on biodiversity in Australia and its territories. Department of Environment and Heritage, Canberra

Crespin SJ, Barahona-Segovia RM (2020) The risk of rediscovery: fast population decline of the localized endemic Chilean stag beetle *Sclerostomulus nitidus* (Coleoptera: Lucanidae) suggests trade as a threat. Insect Conserv Divers 14:107–116. https://doi.org/10.1111/icad.12445

Deacon C, Samways MJ (2021) A review of the impacts and opportunities for African urban dragonflies. Insects 2921(12):190. https://doi.org/10.3390/insects12030190

Del Toro I, Silva RR, Ellison AM (2015) Predicted impacts of climatic change on ant functional diversity and distributions in eastern North American forests. Divers Distrib 21:781–791. https://doi.org/10.1111/ddi.12331

Dennis RLH (2010) A resource-based habitat view for conservation. Butterflies in the British landscape. Wiley-Blackwell, Chichester

Dennis RLH, Shreeve TG, van Dyck H (2006) Habitats and resources: the need for a resource-based definition to conserve butterflies. Biodivers Conserv 15:1943–1966

Diamond SE, Bellino G, Deme GD (2023) Urban insect bioarks of the 21st century. Curr Opin Insect Sci 57:101028. https://doi.org/10.1016/j.cois.2023.101028

Driessen MM (2010) Enhancing conservation of the Tasmanian glow-worm, *Arachnocampa tasmaniensis* Ferguson (Diptera: Keroplatidae) by monitoring seasonal changes in light displays and life stages. J Insect Conserv 14:65–75

Elkinton JS, Parry D, Boettner G (2006) Implicating an introduced generalist parasitoid in the invasive browntail moth's enigmatic demise. Ecology 87:2664–2672

Englund RA, Polhemus DA (2001) Evaluating the effects of introduced rainbow trout (*Oncorhynchus mykiss*) on native stream insects on Kauai Island, Hawaii. J Insect Conserv 5:265–281

Feit B, Gordon CE, Webb JK, Jessop TS, Laffan SW (et al.) (2018) Invasive cane toads might initiate cascades of direct and indirect effects in a terrestrial ecosystem. Biol Invas 20: 1833–1847. https://doi.org/10.1007/s10530-018-1665-8

Fonseca CR (2009) The silent mass extinction of insect herbivores in biodiversity hotspots. Conserv Biol 25:1507–1515. https://doi.org/10.1111/j.1523-1739.2009.01327.x

Forister ML, Pelton EM, Black SH (2019) Declines in insect abundance and diversity; we know enough to act now. Conserv Sci Pract 2019:e80. https://doi.org/10.1111/csp2.80

Fox R (2013) The decline of moths in Great Britain: a review of possible causes. Insect Conserv Divers 6:5–19

Garrard GE, Williams NSG, Mata L, Thomas J, Bekessy SA (2018) Biodiversity sensitive urban design. Conserv Lett 11(2):1–10. https://doi.org/10.1111/conl.12411

Gazzea E, Battisti A, Marini L (2024) Strategies and barriers to reconcile pest management with insect conservation in temperate and boreal forests. Curr For Reps 10(2):103–118. https://doi.org/10.1007/s40725-024-00215-7

Geslin B, Gacher S, Deschamps-Cottin M, Flacher F, Ignace B (et al.) (2020) Bee hotels host a high abundance of exotic bees in an urban context, Acta Oecol 105 (2020) 103556. https://doi.org/10.1016/j.actao.2020.103556

Gibb H, Grubb JJ, Black D, Porch N, Decker O (et al.) (2023) Rainforest litter invertebrates decimated by high severity burns during Australia's gigafires. Austral Ecol 48(7): 1328–1343. https://doi.org/10.1111/aec.13375

Giersch JJ, Hotaling S, Kovach RP, Jones LA, Muhlfeld CC (2017) Climate-induced glacier and snow loss imperils alpine stream insects. Glob Change Biol 23:2577–2589. https://doi.org/10.1111/gcb.13565

Gippet JMW, Bertelsmeier C (2021) Invasiveness is linked to greater commercial success in the global pet trade. Proc Nat Acad Sci 118(14):e2016337118. https://doi.org/10.1073/pnas.2016337118

Gippet JMW, Sherpa Z, Bertelsmeier C (2022) Reliability of social media data in monitoring the global pet trade in ants. Conserv Biol 37:e13994. https://doi.org/10.1111/cobi.140141

Goulson D (2019) The insect apocalypse, and why it matters. Curr Biol 9:R967–R971

Graham KK, Milbrath MO, Killewald M, Soehnlen A, Zhang Y, Isaacs R (2023) Identity and diversity of pollens collected by two managed bee species while in blueberry fields for pollination. Environ Entomol 52:907–917. https://doi.org/10.1093/ee/nvad072

Hall DM, Camilo GR, Tonietto RK, Ollerton J, Ahme K (et al.) (2017) The city as a refuge for insect pollinators. Conserv Biol 31: 24–29. https://doi.org/10.1111/cobi.12840

Harris JE, Rodenhouse NL, Holmes RT (2019) Decline in beetle abundance and diversity in an intact temperate forest linked to climate warming. Biol Conserv 240:108219. https://doi.org/10.1016/j.biocon.2019.108219

Harrison T, Gibbs J, Winfree R (2019) Anthropogenic landscapes support fewer rare bee species. Landsc Ecol 54:967–978

Harvey JA, Tougeron K, Gols R, Heinen R, Abarca M (et al.) (2022) Scientists' warning on climate change and insects. Ecol Monogr 2023; 93:e1553. https://doi.org/10.1002/ecm.1553

Hickling R, Roy DB, Hill JK, Thomas CD (2005) A northward shift of range margins in British Odonata. Glob Chang Biol 11:502–506

Hingston AB, McQuillan PB (1999) Displacement of Tasmanian native megachilid bees by the recently introduced bumblebee *Bombus terrestris* (Linnaeus, 1758) (Hymenoptera: Apidae). Aust J Zool 47:59–65

Hladik ML, Corsi SR, Kolpin DW, Baldwin AK, Blackwell BR, Cavallin JE (2018) Year-round presence of neonicotinoid insecticides in tributaries to the Great Lakes, USA. Environ Poll 235:1022–1029. https://doi.org/10.1016/j.envpol.2018.01.013

Hoffmann B (2014) Quantification of supercolonial traits in the yellow crazy ant, *Anoplolepis gracilipes*. J Insect Sci 14(250):25. https://doi.org/10.1093/jis/14.1.25

Hopping KA, Chignell SM, Lambin EF (2018) The demise of caterpillar fungus in the Himalayan region due to climate changes and overharvesting. Proc Nat Acad Sci 115(45):11489–11494. https://doi.org/10.1073/pnas.1811591115

Horak J (2018) The role of urban environments for saproxylic insects. In: Ulyshen MF (ed) Saproxylic insects. Zool Monogr 1, pp 835–846. https://doi.org/10.1007/978-3-319-75937-1_24

Howarth FG (1991) Environmental impacts of classical biological control. Annu Rev Entomol 36:485–509. https://doi.org/10.1146/annurev.en.36.010191-002413

Hunter MR, Hunter MD (2008) Designing for conservation of insects in the built environment. Insect Conserv Divers 1:189–196. https://doi.org/10.1111/j.1752-4598.2008.00024.x

Janzen DH, Hallwachs W (2019) Perspective: where might be many tropical insects? Biol Conserv 233:102–108. https://doi.org/10.1016/j.biocon.2019.02.030

Jung J-M, Jung S, Ahmed MR, Cho B-K, Lee W-H (2017) Invasion risk of the yellow crazy ant (*Anoplolepis gracilipes*) under the representative concentration pathways 8.5 climate changes scenario in South Korea. J As-Pac Biodiv 10:548–554. https://doi.org/10.1016/j.japb.2017.08.004

Karavin M (2024) Bioaccumulation of metals in some Auchenorrhyncha (Insecta: Hemiptera) species in cherry orchards near motorway and their usage as biomonitor for metal pollution. Bull Environ Cont Toxic 112:26. https://doi.org/10.1007/s00128-023-03843-0

Kehoe R, Frago E, Sanders D (2020) Cascading extinctions as a hidden driver of insect decline. Ecol Entomol 46(4):743–756. https://doi.org/10.1111/een.129985

Koh LP, Sodhi NS, Brook BW (2004) Co-extinctions of tropical butterflies and their hostplants. Biotropica 36:272–274. https://doi.org/10.1111/j.1744-7429

Kotze DJ, Lowe EC, MacIvor JS, Ossola A, Norton BA (et al.) (2022) Urban forest invertebrates: how they shape and respond to the urban environment. Urban Ecosyst 25: 1589–1609. https://doi.org/10.1007/s11252-022-01240-9

Labadessa R, Ancillotto L (2023) Small but irreplaceable: the conservation value of landscape remnants for urban plant diversity. J Environ Manag 339:117907. https://doi.org/10.1016/j.envman.2023.117907

Lassaline CR, Stringham OC, Moncayo S, Toomes A, Cassey JP (2023) Untangling the web: dynamics of Australia's online terrestrial invertebrate trade. Austral Entomol 2023:1–15. https://doi.org/10.1111/aen.12662

Lester PJ (2005) Determinants for the successful establishment of exotic ants in New Zealand. Divers Distrib 11:279–288. https://doi.org/10.1111/j.1366-9516.2005.00169.x

Lester PJ, Beggs JR (2019) Invasion success and management strategies for social *Vespula* wasps. Annu Rev Entomol 64:51–71. https://doi.org/10.1146/annurev.ento.011118.111812

Lester PJ, Keall JP (2005) The apparent establishment and subsequent eradication of the Australian giant bulldog ant *Myrmecia brevinoda* Forel (Hymenoptera: Formicidae) in New Zealand. NZ J Zool 32:353–358. https://doi.org/10.1080/03014223.2005.9518423

Lindemann-Matthies P, Kohler K (2019) Naturalized versus traditional school grounds: which elements do students prefer and why? Urb For Urb Green 46:126475. https://doi.org/10.1916/j.ufug.2019.126475

Losey JE, Chen C, Davis AE, Deitsch JF, Gertin JG, Gorneau JA (et al.) (2022) Insects and spiders on the web: monitoring and mitigating online exploitation of species and services. Glob Ecol Conserv 36, e02098. https://doi.org/10.1016/j.gecco.2022.e02098

Lovett DH, Carr DE (2023) The potential for elevated soil salinity to enhance the ecological trap effect of roadside pollinator habitat. J Insect Conserv 28(1):103–111. https://doi.org/10.1007/s10841-023-00526-3

Mallinger RE, Gaines-Day HR, Gratton C (2017) Do managed bees have negative effects on wild bees?: a systematic review of the literature. PLoS One 12(12):e0189268. https://doi.org/10.1371/journal.pone.0189268

Mason SC Jr, Shirey V, Waite ES, Gallagher MR, Skowronski NS (2023) Exploring prescribed fire severity effects on ground beetle (Coleoptera: Carabidae) taxonomic and functional community composition. Fire 6:366. https://doi.org/10.3390/fire6090366

McGeoch MA, McGrannachan C, O'Connor R, Clarke D, Palmer D (et al.) (2020) Invasive insects: risks, and pathways project. Preliminary results and biosecurity implications. Monash University and Invasive Species Council, Victoria

Merritt DJ, Clarke AK (2013) The impact of cave lighting on the bioluminescent display of the Tasmanian glow-worm *Arachnocampa tasmaniensis*. J Insect Conserv 17:147–153. https://doi.org/10.1007/s10841-012-94903-0

Mola JM, Hemberger J, Kochanski J, Richardson LL, Pearse IS (2021) The importance of forests in bumble bee biology and conservation. Bioscience 71(12):1234–1127. https://doi.org/10.1093/biosci/biab12

Morales CL, Arbetman M, Cameron SA, Aizen MA (2013) Rapid ecological replacement of a native bumble bee by invasive species. Front Ecol Environ 11:529–534. https://doi.org/10.18909/120321

Morales CL, Montalva J, Arbetman MP, Aizen MA, Martins AC, Silva DP (2022) Does climate change influence the current and future projected distribution of an endangered species? The case of the southernmost bumblebee in the world. J Insect Conserv 2002:257–269. https://doi.org/10.1007/s10841-027-00384-5

New TR (2005) Who wants a bull ant for a neighbour? J Insect Conserv 9:71–73

New TR (2014) Insects, fire and conservation. Springer, Cham

New TR (2015) Insect conservation and urban environments. Springer, Dordrecht

New TR (2018) Forests and insect conservation in Australia. Springer, Cham

Peter A, Seress G, Sandor K, Vincze E, Klucsik LP (2020) The effects of artificial light at night on the biomass of caterpillar feeding in urban tree canopies. Urban Ecosyst 323:1311–1319. https://doi.org/10.1007/s11252-020-00999-2

Platts PJ, Mason SC, Palmer G, Hill JK, Oliver TH (et al.) (2019) Habitat availability explains variation in climate-driven range shifts across multiple taxonomic groups. Sci Rep 920190 9:15039. https://doi.org/10.1038/s41598-019-51582-2

Rastogi N (2011) Provisioning services from ants: food and pharmaceuticals. Asian Myrmecol 4:103–120

Ropars L, Dajoz I, Fontaine C, Muratet A, Geslin D (2019) Wild pollinator activity negatively related to honeybee colony densities in urban context. PLoS One 14:e0222316

Rusch A, Bommarco R, Ekbom B (2017) Conservation biological control in agricultural landscapes. Adv Bot Res 81:333–360. https://doi.org/10.1016/bs.abr.2016.11.001

Salafsky N, Salzer D, Stattersfield AJ, Hilton-Taylor C, Neugarten R (et al.) (2008) A standard lexicon for biodiversity conservation: unified classifications of threats and actions. Conserv Biol 22: 897–911

Samways MJ (1988) Classical biological control and insect conservation: are they compatible? Environ Conserv 15:349–354

Samways MJ, Sharratt NJ (2010) Recovery of endemic dragonflies after removal of alien trees. Conserv Biol 24:267–277. https://doi.org/10.1111/j.1523-1739.2009.01427.x

Sands DPA, New TR (2013) Conservation of the Richmond birdwing butterfly in Australia. Springer, Dordrecht

Saunders ME, Barton PS, Bickerstaff JRM, Frost L, Latty T (et al.) (2021) Limited understanding of bushfire impacts on Australian invertebrates. Insect Conserv Divers 14.3: 285–293 https://doi.org/10.1111/icad.12493

Silva LGM, Doyle KE, Duffy D, Humphries P, Horta A, Baumgartner LJ (2020) Mortality events resulting from Australia's catastrophic fires threaten aquatic biota. Glob Change Biol 26:5345–5350. https://doi.org/10.1111/gcb.15282

Spagarino C, Pastur GM, Peri PL (2001) Changes in *Nothofagus pumilio* forest biodiversity during the forest management cycle. 1. Insects Biodiv Conserv 10:2077–2092

Spradbery JP (1988) The European wasp in Australia: present status and future prospects. Proc Sydney Allergen Group 6:78–86

Theodorou P, Radzeviciute R, Lentenu G, Kahnt B, Husemann M (et al.) (2020) Urban areas as hotspots for bees and pollination but not a panacea for all insects. Nat Commun 11:576. https://doi.org/10.1038/s41467-020-14496-6

Threlfall CG, Walker K, Williams NSG, Hahs A, Mata L et al (2015) The conservation value of urban green space habitats for Australian native bee communities. Biol Conserv 187:240–248. https://doi.org/10.1016/biocon.2015/05.003

Tscharntke T (2021) Disrupting plant-pollinator systems endangers food security. One Earth 4(9):1217–1219. https://doi.org/10.1016/oneear.2021.08.022

Tscharntke T, Tylianakis J, Rand TA, Didham RK, Fahrig L (et al.) (2012) Landscape moderation of biodiversity patterns and processes—eight hypotheses. Biol Rev 87: 661–685. https://doi.org/10.1111/j.1469-185X.2011.00216.x

Vanderstock A, Grandi-Nagashiro C, Kudo G, Latty T, Nakamura S (et al.) (2022) For the love of insects; gardening grows positive emotions (biophilia) towards invertebrates. J Insect Conserv 26:751–762. https://doi.org/10.1007/s10841-022-00419-x

Vanderwoude C, Elson-Harris M, Hargreaves JR, Harris EJ, Plowman KP (2003) An overview of the red imported fire ant (*Solenopsis invicta*) eradication plan for Australia. Rec S A Mus, Monogr series 7:111–116

Wagner DL, Fox R, Salcido DM, Dyer LA (2021) A window to the world of global insect declines: moth biodiversity trends are complex and heterogeneous. Proc Nat Acad Sci 118(2):e2002549117

Wang Z, Da W, Negi CS, Ghimire PL, Wangdi K (et al.) (2022) Profiling, monitoring and conserving caterpillar fungus in the Himalayan region using anchored hybrid enrichment markers. Proc Biol Sci 289(1973): 20212650. https://doi.org/10.1098/rspb.2021.2650

Wang Z, Zeng J, Ran H, Meng W, Zhou S (et al.) (2023) Monitoring the online ant trade reveals high biological invasion risk. Biol Conserv 282, 110038. https://doi.org/10.1016/j.biocon.2023.110038

Woinarski JCZ, Chapple DG, Garnett ST, Legge SM, Lintermans M, Scheele BC (2023) Few havens for threatened Australian animal taxa that are highly susceptible to introduced and problematical native species. Biodivers Conserv 33(1):305–331. https://doi.org/10.1007/s10531-023-02750-4

Wood TJ, Goulson D (2017) The environmental risks of neonicotinoid pesticides: a review of the evidence post 2013. Environ Sci Poll Res Int 24:7285–17325. https://doi.org/10.1007/s11356-017-9240-x

Woolley L-A, Murphy BP, Geyle HM, Legge SM, Palmer RA (et al.) (2020) Introduced cats eating a continental fauna: invertebrate consumption by feral cats (*Felis catus*) in Australia. Wildl Res 47: 610–623. https://doi.org/10.1071/WR19197

Yang Z-L (2020) *Ophiocordyceps sinensis* (amended version of 2020 assessment). The IUCN red list of threatened species 2020: eT58514773A179197748.en

Yen A, Butcher RJ (1997) An overview of the conservation of non-marine invertebrates in Australia. Environment Australia, Canberra

Young MR, Barbour DA (2004) Conserving the New Forest burnet (*Zygaena viciae* [Denis and Schiffermueller]) in Scotland; responses to grazing reduction and consequent vegetation changes. J Insect Conserv 8:137–148

Chapter 4
Consequences of Insect Declines and Loss

4.1 Introduction

The wide ecological importance of insects comes to the fore when the consequences of their loss are evaluated and the various 'ecological services' they supply are eroded. Insects play pivotal roles in sustaining ecological functions. Largely unseen, sometimes difficult to appreciate and to value in human terms, the loss of insects equates to loss or erosion of many of the processes people ignore or take for granted in the natural world. The participation of insects in ecological interactions is the very foundation for the continuity of most communities and is essential for those communities to persist. As Memmott et al. (2007) put it, 'Ecological interactions are the currency of ecosystem services'.

Landscape disturbances affecting extent, fragmentation, and condition of any insect habitat may change the balances that maintain such interactions. Anthropogenic disturbances can influence the structure of communities and ecosystems, with the intensity of disturbance affecting the resilience and extent of change. The variety of their impacts, discussed in the previous chapter, extend over both temporal and spatial scales, with the biological responses strongly context-dependent—but with changes amongst the insect fauna often having considerable functional importance.

4.2 Declines of Ecological Integrity

Ecological services or 'ecosystem services' imply the processes needed to sustain ecosystems and range from particular functions undertaken by insects to the disruptions that occur when complex interaction networks are lost, or community composition changes. The services collectively bridge the intrinsic values of natural resources with human-devised socioeconomic 'monetised' systems (Millenium

T. R. New, *Insect Conservation in Australia: Why and How*,
https://doi.org/10.1007/978-3-031-66631-5_4

Ecosystem Assessment 2005). Insects (and other invertebrates) are central players in both the above-ground herbivory-based networks (sometimes referred to as the 'green food webs') and the litter, ground-underground decomposition networks ('brown food webs'). Their full significance and variety of impacts in most of these, however, is unknown. Ecosystem services are commonly adjudged by the benefits to humanity that emanate from ecosystems. These benefits fall into several categories, as 'provisioning' (the supply of products from nature, such as use of insects as food and feed, and biomedical uses), 'supporting' (the basis for production of ecosystem services, such as decomposition and seed dispersal), 'regulating' (the regulation of ecosystem processes, such as pollination, pest control, and soil fertility), and 'cultural' (non-material and aesthetic services to human society, such as bioindicators, recreation [literature, films]) (p. 86), and all are susceptible to loss should the insects involved in maintaining them decline. More familiar to many people are the 'disservices' associated with pest damage or nuisance, by which insect influences on welfare manifest unfavourably, but demonstrate their central ecological significance and influences.

More broadly, encompassing beneficial, undesirable, and unseen impacts, invertebrates have been described as 'webmasters' in ecological systems (Coleman and Hendrix 2000) to convey their central influencing roles and consequences of change. Many years ago, Ehrlich and Ehrlich (1981) likened progressive extinctions in nature to a flying aircraft—as individual rivets dropped out of the wings, the plane would continue to fly until a stage where the accumulated losses resulted in a catastrophic crash: loss of species from an ecosystem may reflect a similar process. It is difficult to embrace the full significance of the 'ecological services' provided by insects, but their collective economic value is enormous (Chap. 6), and their loss is far-reaching. The range of ecological services on which humanity depends can in part be provided by 'domestic' insects—the honeybee (*Apis mellifera*) is a globally vital crop pollinator on which many production systems (and livelihoods of apiarists) depend—but the roles of many wild insects in sustaining the ecosystems they inhabit are unknown, and perhaps effectively undervalued. The consequences of their loss and the subsequent flow or extinction cascades are likewise imponderable. The examples and contexts noted below indicate the variety of concerns raised, from a perspective of human wellbeing. Most studies undertaken to evaluate the values of insects in providing ecological services have been undertaken in temperate region ecosystems, with a strong ecological bias (to terrestrial ecosystems) and taxonomic bias (to Hymenoptera, Coleoptera, and less to Diptera and Lepidoptera). It is not surprising that these 'megadiverse' orders have received most attention, as their obvious richness, ecological variety, and abundance directly imply or suggest ecological significance. However, aquatic insects are no less significant, with aquatic Diptera regarded as major drivers of ecosystem processes (Adler and Courtney 2019). The variety of their services offered to human society 'is limited only by the creativity of investigators' (Adler and Courtney 2019).

The most familiar ecosystem services essential to people are pollination, decomposition, seed dispersal, biological control of pests, and herbivory. They reflect the wide range of feeding habits and ecological interactions in which different insects participate. All operate (and are essential) in both natural and altered environments. The loss of pollinators is perhaps the most widely advanced concern, largely because human food supplies are affected directly as bees and other pollinating insects decline and crop production is lost and people appreciate the seriousness of this. Global declines in pollinators affect reproduction of 90% of wild plants and the yields of about 85% of crop production (quoted by Tscharntke 2021). Pollinators have become a major flagship theme for insect conservation (p. 77). Indeed, Sumner et al. (2018) stated that insect ecosystem services other than pollination are relatively poorly studied.

In short, bees, wasps, flies, butterflies and moths, and beetles can all contribute to pollination needs in different contexts, many plants depend on some form of insect pollination, and many of the details of specialised insect pollinator–plant associations are still unclear; loss of pollinators has serious implications for natural vegetation, in addition to the more applied needs for crop success. Even for the relatively well-studied pollinators, however, the complexity of many interaction webs is still obscure—Memmott et al. (2007) noted that recorded pollination networks were then without exception 'visitation networks', without proof of those visitors being actual pollinators. Those networks are also dynamic—as floral composition changes from climatic or other influences, so may the range of functional pollinators.

In extreme cases, lack of insect pollinators in commercial production systems may demand laborious hand pollinating of fruit and other crops. Initial alarms flowed from widespread decline of honeybees early in this century. *Apis mellifera* is regarded globally as *the* most important crop pollinator, not least because hives can be shifted to seasonally track crops as pollination needs arise. As one example, almond production in California occupied more than 300,000 ha, more than 80% of global production (at the time of a report by Klein et al. 2012), and honeybee pollination was the only realistic option for sustaining production, with growers relying almost entirely on this. Pollination is needed in February, a time when most native pollinators are not active. Interest in increasing the roles of other pollinating insects became more frequent. Management of field margins to increase nesting sites and sources of nectar for wild bees was recommended, as a practice widespread to increase the 'biodiversity values' of agricultural ecosystems (p. 43).

The roles of insects as decomposers and in nutrient recycling were highlighted by the native dung beetles (Scarabaeoidea) of Australia. Dung beetles break down animal dung, promoting accessible forms of nitrogen that act as fertiliser for vegetation. They are important in nutrient recycling and improving soil quality through nitrogen and carbon regulation and sequestration and, in Doube's words 'have the capacity to provide substantial agricultural and environmental benefits' (Doube 2018). The most familiar image of a dung beetle is that of rolling a ball of dung away from its production site to function as a depository for an egg and food for the developing larva. Other species may tunnel into the dung, bury it in situ, or disperse it more widely. Long co-adapted to associate with the small pellet-like faeces of native marsupials, most Australian dung beetles were unable to break down the large amorphous dung pads produced by imported cattle (introduced on the First Fleet in 1788 and, with sheep, founding Australia's pastoral wealth with subsequent massive land transformations to provide pasture), leading later to serious problems of pasture staling as that dung from large numbers of cattle accumulated and persisted: in many cases, cattle will not feed on foliage contaminated by dung, and rank grasses thrived around pads. The dung is also a breeding substrate for nuisance flies. Confining cattle in yards or feedlots also creates waste management problems: Losey and Vaughan (2006) cited a calculation that each animal can produce more than 9000 Kg (or about 23 m^2) of solid waste a year.

A major biological control programme undertaken over many years through CSIRO sought to import and release dung beetles adapted to such dung, from southern Africa and Europe, and which could thrive in the Australian environment (Tyndale-Biscoe 1990); the long history of this innovative enterprise and the major benefits it brought were discussed by Doube and Marshall (2014). These continue, with commercial beetle-rearing exercises and citizen science participation in distributing beetles, and ever-more sophisticated studies of their ecological interactions. More than 50 species were investigated, and screened for suitability for release and 41 species were eventually released, between 1965 and 1985, leading to establishment of 17 species in New South Wales and four each in Victoria and Tasmania. This enterprise led to several species adapting successfully and fulfilling their desired roles in different parts of the country, with the range of each reflecting climatic conditions, and with considerable restructuring of pasture communities. Multiple species were released at some localities, leading to assemblages of several alien species, in some cases leading to competitive effects. Some species spread strongly and colonised large areas where conditions were suitable (Ridsdill-Smith and Edwards 2011; Doube 2018). Four of these species have been considered invasive in various parts of the world: as expressed generally by Pokhrel et al. (2020) they are 'explosive in terms of multiplication, spread, competition and abundance, and have expanded their distributions'. Others did not spread naturally by themselves, and their effectiveness was extended by translocating field-collected beetles to new sites, often with the aid of citizen scientists (p. 147).

In addition to the successful degradation of dung and pasture 'rescue', secondary ecological services of dung beetle activity can include seed dispersal, nutrient recycling, and suppression of stock parasites and flies. The speculative reasons for needing these beetles flow from loss of Australia's historical array of large herbivores, which were likely to have been accompanied by a suite of large dung beetles—paralleling the current situation in parts of Africa. Megafaunal extinctions around 15,000–40,000 years ago may also have led to loss of such beetles to leave only the smaller species adapted to native marsupial dung (discussed by Doube 2018).

Indirect loss of dung beetles, such as by feeding on faeces containing chemicals used to treat livestock for parasites (p. 29) can disrupt their roles, which may also include bioturbation and effects on plants. As one important example, avermectins used to treat internal parasites of cattle in many parts of the world can leave residues in dung and render it toxic to dung beetles and other insects. Some dung beetles also pollinate flowers, including some with decay-like scent (Nicols et al. 2008).

Pest control is essentially the activity needed to suppress the influences of organisms that are undesirable to human interests (so providing 'disservices') and cause economic or other loss. Needs to reduce pesticide use and avoid non-target impacts of all measures imposed have prompted investigations of alternatives such as biological control, through which the ecological services of insects and others are manipulated for pest suppression, both of native pests and, as with the dung beetles noted above, as importations to counter adverse environmental conditions from alien imports (in that case, cattle) or from an alien pest's native range. 'Pest control' has largely been supplanted by 'Pest management' based increasingly on sound ecological principles and with tactics integrated for greatest benefit and least harm to the environment and non-target organisms.

In nature, the impacts of predators and parasitoids are instrumental in preventing many other insects from reaching numbers that could be 'damaging'. Many of these 'natural enemies' are very prey- or host-specific, and they and more generalised native species are frequently the targets of 'conservation biological control', involving preserving or restoring areas around or near arenas (such as agricultural fields) in which the commodity attacked by pests is produced. Field margins and surrounds such as hedgerows can support such biological control agents over periods when the pest is not present, feeding on alternative food species, and constituting reservoir habitats for these species that also provide sources of nectar and pollen for these agents and other insects. Measures to conserve native natural enemies can overlap with pollinator conservation, by providing greater floristic diversity and promoting connectivity (p. 135).

Some classical biological control programmes against alien weeds or arthropod pests have been highly successful, and attention to careful screening—mandatory for all alien organisms to be considered for release in Australia—minimises the risks of non-target impacts. In general (and ideally), pest management should suppress the unwanted impacts of native and alien 'pests' without disrupting the ecological services that insect diversity supplies.

Many insect herbivores contribute significantly to nutrient recycling by dropping faeces and foliage fragments from vegetation. The large scale of this is often not appreciated, but the population sizes and densities of some North American Lepidoptera whose caterpillars defoliate forest trees have in the past been assessed by catching the rain of their faecal pellets in funnels near the ground (pioneered by Morris 1955, in Canada) and the noise of persistent faecal dropping on to house roofs during outbreaks of Gypsy moth (*Lymantria dispar*) can cause media comments in the eastern United States. The same technique enables estimation of the food supply for caterpillar-feeding birds (Switzerland: Fischbacher et al. 1998).

In addition to recycling of nutrients from living foliage as it is consumed, comminution of dead foliage on the ground can also rely on insects. The moth family Oecophoridae, by far the most species-rich family of Lepidoptera in Australia, includes several hundred species whose larvae feed mainly on dead foliage of eucalypts on the forest floor and which otherwise may take a long time to break down naturally. They and larvae of a group of chrysomelid beetles are regarded as key components of the decomposer guild, as are termites and some other insects.

In aquatic ecosystems, some Diptera and others function as 'bioturbators', that through their burrowing and feeding activities and subsequent defecation mobilise nutrients to become available for other organisms. Densities of some aquatic fly larvae are remarkably high—blackfly (Simuliidae) and mosquito (Culicidae) larvae can reach more than 100,000 larvae/m^2: the latter have been likened in function to earthworms, eating dead organic matter, and defecating to aerate and add nitrogen to their habitat. A study of blackfly larvae in a Swedish river showed that daily transport of faecal pellets past a defined point of the river could reach 429 tons of dry mass—a staggering figure that Adler and Courtney (2019) picturesquely considered approximately equivalent to 6000 elephants defecating (wet weight) into a river each day!

The vast numbers of social insects, most notably ants and termites, and their ecological roles, give them enormous significance in terrestrial ecosystems. Ant and termite richness is greatest in the tropics and attenuates at higher latitudes, with termites largely found in the tropics and subtropics, so with major influences are more restricted. Ants, however, occur as a predominant insect group over temperate regions. Red wood ants (*Formica* spp. of the *Formica rufa* group), for example are forest keystone taxa over much boreal and temperate forests of the northern hemisphere, and valued in control of forest insect pests. Measures to manage forests to prevent further losses reflect this appreciation, and need to harmonise extractive forestry with the survival of particular ant species. For example, it might be necessary to create forest gaps and clearings where canopy closure is disadvantageous. Wood ant declines in Europe have led to some species being listed for protection; that protection is very uneven across European countries (Balzani et al. 2022). These ants, whose specific taxonomy is complex, are also 'umbrella species' because their large nest mounds harbour numerous co-existing myrmecophilic invertebrates, some of which are also listed; some are also obligate mutualists that depend entirely on the ants for their own survival.

More broadly, ants exert impacts 'that are unmatched by most animal clades' (Parker and Kronauer 2021).

The major role of termites, with their biomass dominating many arthropod soil faunas, is as soil ecosystem engineers, a role shared with ants and earthworms, with the outcomes of 'conditioning' soil, such as by rehabilitating degraded soil, improving soil aeration through their nesting activities, and aiding the decomposition of dead organic matter. Termites feed on cellulose in organic matter, in soil, or in dead wood, and removal of forests may diminish those services; they are also an important food source for numerous vertebrate and invertebrate consumers.

Many land-use changes involve conversion of forest or woodland for other purposes, with losses of resident forest termites. In Bignell's (2019) words, this 'often begins a degenerative sequence in which biodiversity, soil fertility, soil ecosystem resilience and soil physical stability all decline'. Termites are thus regarded as vitally important to soil health. Changing soil community dynamics, affecting rates of nitrogen sequestration, are often difficult to estimate—for example, honeydew removal by *Vespula* wasps in New Zealand (p. 50), reducing its input to the ground, may change soil microbial communities from predominantly bacteria-based to fungus-based (Lester and Beggs 2019).

However, despite their notoriety, Elizalde et al. (2020) claimed that many of the ecosystem services provided by social insects are not widely known or valued (Table 4.1), and recognised 13 categories of services. However, the numbers of categories of ecological services recognised differ somewhat across accounts and taxa. At one extreme, a broad literature review (Noriega et al. 2018) enumerated 42 specific roles of insects across the four major categories (Table 4.2). More specifically, Rastogi (2011) summarised the roles of ants (Fig. 4.1), and discussed their relevance to 'provisioning services' through being food and medicine, whilst Elizalde et al. noted the wider roles of social insects in provisioning and that the traditional knowledge of local people is fundamental in evaluating these. Importantly, social insects can strongly influence other species as well as the structure and function of ecological communities. Ants, the most diverse group of social insects, are dominant components of many terrestrial communities, in which their taxonomic and ecological variety has given them a unique role in appraising environmental condition and change. 'Functional groups' amongst ants, many based on trophic roles and exemplified by particular taxonomic segregates, have facilitated interpretation and accelerated research into the significance of these insects in ecosystem functions. Each functional group may be linked with a primary functional role, some of which have been investigated in considerable depth. Seed dispersal is the most widely studied regulating ecosystem process, reflecting wider interests in myrmecochory—a suite of complex co-evolved mutualistic systems in which plant seeds provide an attractive food source (the elaiosome) for ants and which the ants collect and disperse the seed. Myrmecochory has developed multiple times. It is widespread geographically and, as it has been reported for more than 11,000 plant species, may be important in maintaining plant diversity (Del Toro et al. 2012). In some temperate woodlands, ants are reported to disperse 40% of all herbaceous

Table 4.1 Ecosystem services provided by social insects, abbreviated from listing by Elizalde et al. (2020) Major groups of social insects are abbreviated as T (termites), A (ants), B (bees), W (wasps)

Category	Service	Taxa
Provisioning	Pharmaceuticals and medicine	T, A, B, W
	Food	T, A, B, W
	Others (biofuel, silk, wax)	T, A, B
Regulating	Pollination	A, B, W
	Biological control	A, B, W
	Seed dispersal	T, A, B, W
Supporting	Bioturbation	T, A
	Nutrient cycling and decomposition	T, A
	Carbon cycling (C mineralisation)	T, A
	Energy flow	T, A
Cultural (non-material)	Biological indicators	T, A, B, W
	Literature and arts	A, B, W
	Cultural traditions and religion	T, A, B

Table 4.2 The ecosystem services provided by insects, as listed by Noriega et al. (2018), based on literature review of papers published 1956–2016, to illustrate the variety of processes in which insects participate

Category	Services listed
Provisioning services	Alternative nutrition source; Economic Services; Food chain supplementation; Industrial production; Medicine services
Regulating services	Below ground exchange; Carbon absorption; Climate regulation; Control and suppression of pathogens; Counteract climate change; Fungus control; Gastrointestinal parasite control; Greenhouse gas emissions; Habitat genetic diversity; Network services; Pest control; Pollination; Population regulation; Soil fertility regulation; Soil nutrient regulation; Soil nutrients spatial variability; Soil erosion prevention
Supporting services	Biodiversity protection; Decomposition; Dung removal; Hydrological soil properties; Mineralisation; Nutrient accumulation; Nutrient flow; Recycling of matter; Seed dispersal; Soil removal; Soil structure; Soil water infiltration
Cultural services	Bioindicators tool; Conservation tool; Cultural heritage; Education; Recreation services; Religion and spiritual values; Tourism services; Urban quality life

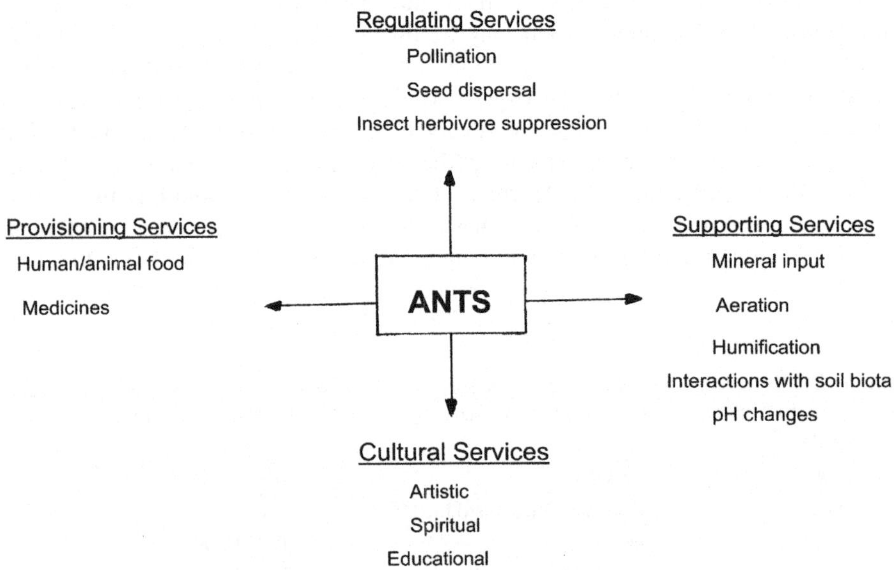

Fig. 4.1 The variety of ecosystem services provided by ants, expressed as the four main categories (provisioning, regulating, supporting, and cultural services) with representative major examples of each (after Rastogi 2011, used under Creative Commons licence CC-BY-4.0)

plants, and myrmecochory is most developed in Australia, eastern North America, and southern Europe.

Many interactions amongst species are complex and their influences on ecological processes are unclear. Several manipulative investigations demonstrate these complexities. Recognising the crucial roles of termites in decomposition in tropical systems, Walker et al. (2022) showed that predation of termites by ants strongly influenced termite activity. They suppressed ant activity (using insecticide baits) in 1-hectare plots in South African savanna, and compared decomposition rates of wood, grass, and dung in treated and untreated plots to show that decomposition rates of these increased by 98%, 74%, and 84%, respectively, with ant suppression, in all nearly doubling decomposition by termites in natural plots. A widespread study (Zanne et al. 2023) also inferred that termite activity is likely to increase with warming temperatures.

Most of the examples noted above are essentially anthropocentric, as showing benefits to humanity from insect activities (and as the epithet 'services' implies), but the vast numbers and diversity of insects give them far wider ecological importance. At least three-quarters of flowering plant species depend on insects for pollination, for example. Estimates for the wider 'animal pollination' (IPBES 2016) were 87.5% of the approximately 308,000 flowering species, suggested to range from 94% in tropical communities to 78% in temperate zone communities. The wealth of wild bees (most of them solitary and unknown to people who regard the honeybee as 'the bee', and that has led to much public concern and conservation advocacy as *Apis mellifera* itself has declined), hoverflies, beetles, moths and others contributing to this include many that are specialised and form essentially mutualistic or obligatory associations with particular plant taxa. Some represent, or are believed to represent, a long history of coevolution to form very specific relationships. Some such insects are referred to as 'keystone species', with ecological influences far greater than their abundance might suggest. The South African fly *Moegistorhyncus longirostris* (Nemestrinidae) has the longest proboscis known in the Diptera, up to 83 mm long (Barraclough and Slotow 2010). It pollinates the long-tubed flowers of at least 20 flower species across several families, several of which are probably pollinated only by this fly. Loss of pollinators has far wider implications than human food security alone, and increased concerns are leading to calls for pollinator conservation in policy and practice—driven mainly by food security issues but with the major documents also presenting the wider needs. An Intergovernmental Science-Policy Platform on Biodiversity and Ecosystem Services report (IPBES, Potts et al. 2016) gave an overview of knowledge of pollinators and their needs, confirming large-scale declines in western Europe and North America and concluding that pollinators (including vertebrates such as hummingbirds and bats) are under increasing threat. That report listed a number of 'key messages' for scientists and policymakers, based on the need for crop pollination and the wider ecological perspective of non-crop needs. Those messages convey the variety of values, status, and trends in pollinators and pollination, the needs and approaches as components of pollinator roles and conservation development, and collectively emphasise the variety of strategic measures needed to ameliorate the equivalent variety of threats to bees and other animal pollinators. The final key measure recognises the need for effective collaboration across sectors in order to bring positive outcomes: it reads, in part, 'Coordinated, collaborative action and knowledge sharing that builds links across sectors (e.g., agriculture and nature conservation), across jurisdictions (e.g., private, Government, not-for-profit), and among levels (e.g., local, national, global) can overcome these challenges and lead to long-term changes that benefit pollinators'.

The report was followed by an independently suggested list (Dicks et al. 2016) of ten policies for governments to pursue to protect pollinators and secure pollination services (Table 4.3). National strategies for pollinator conservation in the United States, England, and France, and subsequent comments on the IPBES assessment all confirm concerns, but recommended policy modifications are (1) sometimes slow to be heeded and (2) unevenly adopted across a region or over divisions within a country. For example, the best scientific information on pollinators supplied for legislative use in Brazil was not rapidly taken up and Hipolito et al. (2021) urged that Brazilian scientists should be consulted more often and participate in proposals for laws related to pollinator conservation.

Later revision of the European policy (European Commission 2023) set out actions needed to reverse the decline of pollinators, under three main categories: (1) improving knowledge of pollinator decline, its causes and consequences; (2) improving pollinator conservation and tackling the causes of their decline; and (3) mobilising society and promoting strategic planning and cooperation at all levels. These relate to targets set earlier to be achieved by 2030, but again emphasise the widespread appreciation of the seriousness of pollinator losses.

Insect declines from pesticides and ALAN diminishes food supply for aerial insectivores, notably bats and birds. Declines amongst these taxa have been reported from several parts of the world and, whilst difficult to confirm, correlations suggest that reduced insect prey supply may be an important contributor. Declines of the iconic Australian Bogong moth (*Agrotis infusa*, Noctuidae) from factors including pesticide uses over their extensive lowland breeding areas, and resulting in only small numbers of adults aestivating in their traditional cave sites in alpine areas, have removed a major food source for the Mountain pygmy possum (*Burramys parvus*). Concern for this endangered marsupial has become the main stated purpose of promoting conservation of the moth.

Table 4.3 The ten 'policies for pollinators' suggested to help governments protect pollinators and secure pollination services: from Dicks et al. (2016), who commented that governments should 'seriously consider' these policies

Raise pesticide regulatory standards
Promote integrated pest management (IPM)
Include indirect and sublethal effects in genetically modified crop risk assessments
Regulate movement of managed pollinators
Develop incentives, such as insurance schemes, to help farmers benefit from ecosystem services instead of agrochemicals
Recognise pollination as an agricultural input in extension services
Support diversified farming systems
Conserve and restore 'green infrastructure' (a network of habitats that pollinators can move between) in agricultural and urban landscapes
Develop long-term monitoring of pollinators and pollination
Fund participatory research on improving yields in organic, diversified, and ecologically intensified farming

The penetrating survey of declines of insects in a region of China by Zhou et al. (2023) (p. 28) illustrated the progressive loss of ecological functions that resulted, through losses of regulation of herbivores as natural enemies declined. Together, 124 individual linkages of 'natural enemy-host' provided a complex trophic network centred on the pest noctuid moth *Helicoverpa armigera*. Overall, during the 18-year study, abundance of natural enemies fell significantly faster than that of their paired herbivore prey/hosts. Crop pest populations declined by 63.7% over this period, compared with natural enemies declining at 80%. The changed 'balance' of trophic units and so of interactions in this agricultural community suggests that parallel changes could occur in many other communities, leading to widespread ecological changes. They are extremely difficult to measure, and the specificity of most of the participants is unknown; nevertheless, Zhou et al.'s study is a salutary warning of possible future trends.

References

Adler PH, Courtney GW (2019) Ecological and societal services of aquatic Diptera. Insects 10:70. https://doi.org/10.3390/insects10030070

Balzani P, Dekoninck W, Feldhaar H, Freitag A, Frizzi F (et al.) (2022) Challenges and a call to action for protecting European red wood ants. Conserv Biol 36 e13959 https://doi.org/10.1111/cobi.13959

Barraclough D, Slotow R (2010) The South African keystone pollinator *Moegistorhynchus longirostris* (Wiedemann, 1819) (Diptera: Nemestrinidae): notes on biology, biogeography and proboscis length variation. Afr Invertebr 51:397–403

Bignell DE (2019) Termite ecology in the first two decades of the 21st century: a review of reviews. Insects 10(3):60. https://doi.org/10.3390/insects10030060

Coleman DC, Hendrix PF (2000) Invertebrates as webmasters in ecosystems. CABI Publishing, Wallingford

Del Toro I, Ribbons RR, Pelini SL (2012) The little things that run the world revisited: a review of ant-mediated ecosystem services and disservices (Hymenoptera: Formicidae). Myrmecol News 17:133–146

Dicks LV, Viana B, Bommarco R, Brosi B, Arizmendi MdC (et al.) (2016) Ten policies for pollinators. Science 354: 975–976. https://doi.org/10.1126/science.aai9226

Doube BM (2018) Ecosystem services provided by dung beetles in Australia. Basic Appl Ecol 26:35–49. https://doi.org/10.1016/j.baae.2017.09.008

Doube BM, Marshall T (2014) Dung down under: dung beetles for Australia. Dung Beetle Solutions Australia, Adelaide

Ehrlich PR, Ehrlich AH (1981) Extinction: the causes and consequences of the disappearance of species. Random House, New York

Elizalde L, Arbetman M, Arnan X, Eggleton P, Leal IR (et al.) (2020) The ecosystem services provided by social insects: traits, management tools and knowledge gaps. Biol Rev 95.5: 1418–1441. https://doi.org/10.1111/brv.12616

European Commission (2023) Communication for the Commission to the European Parliament, the Council, the European Economic and Social Committee and the Committee of the Regions. Revision of the EU Pollinators Initiative. A new deal for pollinators. COM/2023/35 final

Fischbacher M, Naef-Daenzer B, Naef-Daenzer L (1998) Estimating caterpillar density on trees by collection of frass droppings. Ardea 86:121–129

Hipolito J, Coutino J, Mahlmann T, Santana TBR, Magnusson WE (2021) Legislation and pollination: recommendations for policymakers and scientists. Perspect Ecol Conserv 19:1–9. https://doi.org/10.1016/j.pecon.2021.01.003

IPBES (2016) Summary for policymakers of the assessment report of the intergovernmental science-policy platform on biodiversity and ecosystem services on pollinators, pollination and food production. (see Potts et al. 2016, below)

Klein A-M, Brittain C, Hendrix SD, Thorp R, Williams N, Kremen C (2012) Wild pollination services to California almond rely on semi-natural habitat. J Appl Ecol 49:723–732. https://doi.org/10.1111/j.1365-2664.2012.02144.x

Lester PJ, Beggs JR (2019) Invasion success and management strategies for social *Vespula* wasps. Annu Rev Entomol 64:51–71. https://doi.org/10.1146/annurev.ento.011118.111812

Losey JE, Vaughan M (2006) The economic value of ecological services provided by insects. Bioscience 56:311–323

Memmott J, Gibson R, Carvaheiro LG, Henson K, Heleno RH (et al.) (2007) The conservation of ecological interactions. pp. 226–244 in Stewart AJA, New TR, Lewis OT (eds) Insect conservation biology. CAB International, Wallingford

Millenium Ecosystem Assessment (2005) Summary for decision makers. Island Press, Washington, DC

Morris RF (1955) The development of sampling techniques for forest insect defoliators, with particular reference to the spruce budworm. Can J Zool 33:225–294

Nicols E, Spector S, Louzada J, Larsen T, Amezquita S et al (2008) Ecological functions and ecosystem services provided by Scarabaeinae dung beetles. Biol Conserv 141:1461–1474. https://doi.org/10.1016/j.biocon.2008.04.011

Noriega JA, Hortal J, Azcarate FM, Berg MP, Bonada N (et al.) (2018) Research trends in ecosystem services provided by insects. Basic Appl Ecol 26: 8–23. https://doi.org/10.1016/j.baae.2017.09.006

Parker J, Kronauer DJC (2021) How ants shape biodiversity. Curr Biol 31:R1141–R1224

Pokhrel MR, Cairns SC, Andrew NR (2020) Dung beetle species introductions; when an ecosystem service provider transforms into an invasive species. Peerj 8:e98972. https://doi.org/10.7717/peerj.9872

Potts SG, Imperatriz-Fonseca VL, Ngo HT, Biesmeijer JC, Breeze TD (et al.) (eds) (2016) IPBES (2016) summary for policymakers of the assessment report of the intergovernmental science-policy platform on biodiversity and ecosystem services on pollinators, pollination and food production. Zenodo. https://doi.org/10.5281/zenodo.3402857

Rastogi N (2011) Provisioning services from ants: food and pharmaceuticals. Asian Myrmecol 4:103–120

Ridsdill-Smith TJ, Edwards PB (2011) Biological control: ecosystem functions provided by dung beetles. In: Simmons LW, Ridsdill-Smith TJ (eds) Ecology and evolution of dung beetles. John Wiley and Sons, London, pp 245–266

Sumner S, Law G, Cini A (2018) Why we love bees and hate wasps. Ecol Entomol 42:836–845. https://doi.org/10.1111/een.12676

Tscharntke T (2021) Disrupting plant-pollinator systems endangers food security. One Earth 4:1217–1219. https://doi.org/10.1016/oneear.2021.08.022

Tyndale-Biscoe MM (1990) Common dung beetles in pastures of South-Eastern Australia. CSIRO Publishing, Melbourne

Walker AEL, Robertson MP, Eggleton P, Bunney K, Lamb C (et al.) (2022) Indirect control of decomposition by an invertebrate predator. Funct Ecol 36: 2943–2954. https://doi.org/10.1111/1365-2435-14198

Zanne A, Flores-Moreno H, Powell JR, Cornwell WK, Dalling JW (et al.) (2023) Termite sensitivity to temperature affects global wood decay rates. Science 377: 1440–1444. https://doi.org/10.1126/science.abo3856

Zhou Y, Zhang H, Liu D, Khashaveh A, Li Q (et al.) (2023) Long-term insect censuses capture progressive loss of ecosystem functioning in East Asia. Sci Adv 9, eade9241

Chapter 5
Human Perceptions of Insects

5.1 Introduction

The adverse perception of insects by many people is perhaps the most widespread hindrance—even, impediment—to promoting their conservation. As Losey et al. (2022) noted, 'there is often a sociocultural perception that invertebrates are of less importance than vertebrate species' and should thereby be ranked lower for conservation attention and resources. The single best-known supporting slogan for insect conservation is perhaps Wilson's (1987) now famous designation of invertebrates as 'The little things that run the world', a phrase that continues to resonate strongly in public and political advocacy, but there perhaps less so than in the academic world (Donkersley et al. 2022). Likewise, they claimed that more recent catchphrases such as the 'Insect Armageddon' have not persisted strongly after a considerable initial impact. Any such designation that can capture public sympathy or attention to insect roles and fate may contribute to conservation interest and understanding the need. And any such attitudinal transformation is also a key need in future conservation endeavour. The problem relates strongly to how people regard insects, either adversely through phobia, fear, annoyance, actual or potential aesthetic or economic harm, or positively through cultural or aesthetic appreciation, economic gain or—more rarely—acknowledgment of insect values in the biosphere. Many people, in reality, do not 'regard' insects at all: they are simply ignored or dismissed except when they intrude on consciousness or wellbeing though some fear, nuisance or economic impact. In a somewhat different viewpoint, Kellert (1993) suggested that many people have an 'innate learning disposition' reflected in unwillingness to seriously understand invertebrates as a 'low form of life' with no sense of identity and consciousness! As Hall and Martins (2020) put it 'It is striking how invisible insects are to the public'. And, more embracingly, for 'a world that values insects' (Basset and Lamarre 2019), those values must become both apparent and appreciated.

T. R. New, *Insect Conservation in Australia: Why and How*, https://doi.org/10.1007/978-3-031-66631-5_5

Perhaps the most urgent and important tasks facing insect conservationists are to transform public opinion and garner wider support for insect life.

However, that transformation must occur against the reality that many insects are indeed harmful to human wellbeing, and have been so over many centuries—some as vectors of disease, with transmission of malaria by *Anopheles* mosquitos, of a considerable variety of viruses by mosquitos and midges, sleeping sickness by tsetse flies, and others needing constant vigilance and concerns. Likewise, the economic impacts of many agricultural, horticultural, forestry, stored commodity and other pests affect numerous livelihoods and wider human welfare. Consequently, insects are 'probably the costliest animal group to human society' (Bradshaw et al. 2016). Any such harmful insect is an understandable target for suppression, a perspective unlikely to be modified if human health and needs are to be protected.

Any widespread transition from (non-pest) insects being ignored or viewed negatively to them becoming viewed positively—even elevating them to a component of 'biophilia' through changed disposition—requires extensive education and efforts to increase awareness and appreciation of insect life, essentially to transform insect 'biophobia' to 'biophilia'. Awareness and understanding, as the twin avenues to garnering support for insect conservation, both flow essentially from sympathetic perceptions. Cardoso et al. (2011) quoted Janzen's maxim for conservation that 'if you don't know it, you can't love it. If you don't love it, you won't save it'. Most insects remain 'unloved'. Lockwood (2013) explored attitudes to insects in detail, with long-held cultural sentiments, some perhaps innate, from far back in human history the foundation of more diverse modern reactions. The psychology of human-insect relationships is complex. Berenbaum (2008) used the term 'charisma-challenge' to express widespread perceptions of insects, and even being common might lead to some insects seeming less threatening (Guiney and Oberhauser 2008).

5.2 Perception: Good and Bad Insects

Nevertheless, the few 'iconic' insect groups accepted readily by people as in some way 'worthy'—most notably butterflies and dragonflies (sometimes referred to as 'birdwatcher's bugs' because of their diurnal habits, visual appeal and that they can be observed constructively with binoculars or a monocular)—are powerful symbols for promoting information and conservation worth of both themselves and other insects and the variety of insect habitats. They have demonstrated the great variety of insect biology, habitat needs and ecological subtleties amongst a group of predominantly terrestrial herbivores and aquatic predators, respectively. Both groups have become important foci for insect conservation—not simply as iconic and aesthetically appealing, but also because many are sensitive to environmental changes and their responses can be monitored and evaluated. Concerns over Lepidoptera and Odonata continue to be a paramount stimulus for wider insect conservation. In essence, they have become widespread surrogates for much other insect life, and 'umbrellas' for wider insect conservation. In general, the most appealing insects to many people are large, colourful, diurnal and perceived as harmless. Appeal may be enhanced by some structural or behavioural 'oddity'; appreciation of giant cockroaches and phasmids as pets, for example, reflects their unusual nature, as well as ease of maintenance.

The legacy of 'liking' insects for aesthetic, cultural or other reasons is also the foundation of interest in collecting, an activity that can expand to include 'insect substitutes' rather than actual insects—for example, insects on postage stamps. Entomophilately is a popular cultural aspect of thematic stamp collecting, and now includes more than 20,000 'insect stamps' from many countries (Nazari 2021, noted 20,341 stamps up to 2020, with 551 of those depicting arachnids), and ranging from commemoratives of events, conferences and portraits of notable entomologists to pictures of many different insects, some in scientific contexts. Numerous orders are included, with Lepidoptera and Odonata predominant and several other diverse orders (Coleoptera, Hymenoptera, Diptera) under-represented. However, taxonomy is not always reliable: Nazari (2021) found that 30% of the stamps contained errors such as misidentifications, incorrect spelling, incorrect common name, and others that collectively limited the values of the stamps as educational aids. As well as appeal to philatelists, stamps are an economic item purchased for postage and, as Leather (2015) noted, they play a dual economic and cultural role. Other pastimes, such as films (p. 86) and video games focused on insects can have a strong educational/insect awareness component.

'Disliking' insects, perhaps far more widespread than liking them, is complex to interpret and can have strong societal influences.

Entomophobia—usually defined as an excessive and irrational fear of insects—is daunting to sufferers, and can commonly manifest as 'delusional parasitosis', whereby an individual believes the anxiety, fear, increased heart rate, and other symptoms experienced are due to insect infestation. The belief may sometimes originate from an experience such as a bee sting or other disturbance and can persist for a considerable time: children may learn the phobia from parents, for example. Medical examinations invariably fail to find any trace of culpable insects—or, often, of any other cause. Either all insects or particular insect groups may be the instigator, and treatment involves sympathetic and empathetic counselling and accompanying medication.

A somewhat different 'phobia' is the dislike and distrust of entomophagy, which Yen (2009a, b) and others have highlighted as a major impediment to increasing adoption and promotion of insects as human food. Whilst both terrestrial and aquatic insects are a staple in many parts of the world (p. 98), considerable reluctance to eat them—even by people who readily devour other invertebrates such as some molluscs and crustaceans—remains in 'western society'.

Nevertheless, the image of insects being pests remains prevalent over much of society, and can lead to undiscerning condemnation of insect life—but still with acknowledgement that some insects are indeed 'good'.

Perceptions of insects as 'good' or 'bad' by people have a strong taxonomic component or bias; simplistically, for example, many people 'like' butterflies but dislike moths, and like bees but dislike and fear wasps., as revealed in several recent surveys. The contrasting public image of 'locusts' as major pests and closely related (even, conspecific) 'grasshoppers' as innocuous and aesthetically pleasing harbingers of summer tranquillity can involve taxa that even specialists find difficult to distinguish—or, even, the same species in different behavioural phases. The first (Lepidoptera) in part reflects traditional cultural appreciations and contrasted perception, based on generalities that mask the full variety present (such as 'butterflies are brightly coloured, diurnal, not pests or harmful, and are symbols of wellbeing; moths are drab, nocturnal, harmful pests, with little redeeming image'). For the second example, Sumner et al. (2018) suggested that dislike of wasps is fostered largely through the most familiar stinging wasps (Vespidae) that people learn to recognise, avoid and fear as part of a wider evolutionary move toward avoiding organisms that can sting, bite, kill or spread disease. Dislike, fear or revulsion of such insects is a strong barrier to gaining conservation sympathy. Whilst conservation of bees is supported widely for sustaining pollination services, conservation of social wasps, important as predators but also representing a far wider range of services flowing from aculeate wasps (Brock et al. 2021), is far less easily supported, with surveys of urban community gardeners in Germany, for example, revealing strong negative perceptions of social wasps, driven in part by lack of knowledge of their roles (Schmack et al. 2024). The notoriety of some invasive Vespidae as threats to native insects and the balance of natural ecosystems (p. 50)—as well as danger to people—can outweigh appreciation of their values in more natural environments, in which the at least 30,000 species of aculeate wasps may have considerable impacts on arthropod populations (Brock et al. 2021), with their roles in natural pest management largely overlooked. Wider public support for bees is in part because people appreciate the need for pollination—and like honey! The apiculture industry, combining pollination services with commercial supply of bee products, continues to expand. These values outweigh the fears of stinging, whilst wasps are not seen to have any such redeeming benefits. However, large numbers or concentrations of feral honeybees in inland Australia can become a concern for tourists and reduce their otherwise positive image by fears of being stung; in hot dry weather large numbers of bees seeking moisture around the few water sources (tanks, toilets) used by campers in Wyperfeld National Park, northwest Victoria, for example prompt posting of warning notices for visitors to take care.

The ecological services provided by insects are often overlooked by people. It has been suggested that 'fearing' insects may be intrinsic, whereas 'loving' insects may have to be learned through positive interactions (Samways 2018), and that a general lack of knowledge of insects also leads to failure to distinguish the relatively few pests or harmful taxa from the great innocuous or beneficial majority, thus enhancing negative attitudes (Fukano and Soga 2021). The major human negative emotions towards insects broadly comprise 'fear' and 'disgust' and, intriguingly, these feelings may be linked with urbanisation—people then see more 'bugs' indoors and, with decline of nature-based experiences, inability to distinguish between them is also increased (Fukano and Soga 2021); people in urban areas may have particularly negative attitudes toward insects, simply because frequent encounters reinforce a negative image. Conversely, outdoor activities such as home gardening can help lessen fear of insects (Vanderstock et al. 2022).

The roles of vast numbers of swarming locusts as herbivorous pests flow from prehistory and biblical times, and persist. Those swarms develop at intervals among species that are normally solitary grasshoppers, but under particular environmental conditions nymphs undergo physiological and behavioural changes causing them to become gregarious and form 'bands' of hoppers that move around and strip vegetation, later becoming flying adults: the term 'locusts' refers to this gregarious phase, rather than any taxonomic distinction. A large locust swarm can extend over thousands of square kilometres with insect densities of up to about 80 million/$Km^{2,}$ and may persist for several years. As the targets of massive control programmes (including aerial applications of pesticides) against hoppers and adults, some locusts are among the most damaging of all agricultural insect pests. Yet some are congeneric and very similar in appearance to Orthoptera that are demonstrably rare and are in some cases threatened: some, indeed, have become extinct (Samways and Lockwood 1998), and others are the focus of active conservation measures. Harmonising needs to suppress severe pests and conserve rare congeneric (or other) species can become complex (Samways and Lockwood described it as a 'paradox'), but these orthopterans again illustrate the contrasting values in human perception—and the difficulty of gaining consensus and conservation sympathy.

5.3 Improving the Images of Insects

'Valuing insects' reflects appreciation of their positive or materialistic attributes or benefits. Many of these are embedded in early culture, economy, and human life. Honey gathering from nests of wild bees started around 7000 BC as perhaps the first tangible positive value of insects to humanity. Sericulture based on the 'silkworm' (*Bombyx mori*) in China had become an important peasant industry by 4000–3000 BC. The 'lac insects' (Hemiptera: Kerriidae) have long been cultivated commercially, with their dye, resin, and wax products having many industrial applications (Bashir et al. 2022). Cultivated over much of southeast Asia, and with the greatest production in India, the lac industry continues to thrive and provide numerous livelihoods. Even locusts were, and are, valued as food and considered a delicacy in many countries. Likewise, the brood of social wasps is harvested for human food in many places (Brock et al. 2021). Cultural images of wasps as 'villains' in a number of alarmist/science fiction films (below) and literature can broaden prejudice. Appreciating that the variety of cultural services provided by insects collectively includes services related to art, recreation, and the development of traditional belief systems (Duffus et al. 2021)—but for all of which values are difficult to quantify in conventional terms that can increase their public appeal—helps shape the relationships between people and insects.

Since 1984, the University of Illinois at Urbana-Champagne has hosted an annual Insect Fear Film Festival, as an initiative by Prof May Berenbaum. Hymenoptera is the most frequently depicted insect group in 'Fear Films' (Berenbaum and Leskovsky 2009). Each Festival emphasises a particular theme; for example the 39th Festival, 2022, was titled 'Venom' and its scope, in the words of the advertising notice and in three groups of presentations, was 'first, stings as exaggerated comic plot elements; next, reactions to stings, including anaphylaxis, as humorous, dramatic or horrific highlights; and, finally, shorts revealing the scientific reality of venom and venomous insects, including potential beneficial uses of arthropod venom'. The 40th Festival (in 2023) featured 'Living Fossils'. Such combinations of entertainment and information are a powerful and, in this case, highly successful way to modify perceptions of insects.

More widely, various other forms of 'insect festival' are becoming more numerous. They range from a focus on particular insect groups, some—such as fireflies or dragonflies—of tourist or cultural interests to more general celebrations of insect life. They link with 'insect zoos', in which series of living invertebrates are shown, or taken into schools for children to handle and help overcome their fears as insect features are explained by keepers. Handling (in Australia) such animals as the Giant burrowing cockroach (*Macropanesthia*), Giant prickly stick insect (*Extatosoma tiaratum*), Lord Howe Island stick insect (*Dryococelus australis*), Emperor gum moths (*Opodiphthera* spp., with the 'furry abdomen' of these and other large silk moths sometimes believed to be an attractive feature—as for bumblebees elsewhere in the world) and others can create lasting and positive memories.

Fear of stinging Hymenoptera is justifiably widespread, as leading to severe pain and, more rarely, anaphylactic shock. Thus, in the United States, Golden (2015) estimated that less than 1% of children and 3% of adults have systemic allergic reactions to insect stings, and others have unusually large local inflammation. Anaphylaxis can lead to death: several deaths in Australia have been caused by stings of an endemic Jumper ant, *Myrmecia pilosula*. Less authoritatively but nevertheless contributing to alarm, 'killer bees' and allied themes occur repeatedly in films (Berenbaum and Leskovsky 2009) and science fiction.

Many films, although reality is distorted by animation, sensationalism, or 'horror', deal with ecological issues such as impacts of pesticides and pollution, disruption of food webs, or impacts of alien species. Likewise, references to insects in books span many levels of reality and imagination. Leather (2015) listed examples ranging from Aesop's fables, through the young persons' classic 'The Very Hungry Caterpillar' by Eric Carle (Carle 1969), to adult novels and science fiction, collectively sufficient to cause strong reactions from readers. Content of comics is also widely read—Leather noted that the Marvel and DC comics have together based 232 'superheroes' and villains on insects or with insect-like characteristics, as discussed in more detail by Da Silva et al. (2014); still others are based on arachnids, and their numbers continue to increase. Insects appear in many song lyrics and music.

The 'Schmidt Pain Index' ranked the pain level of Hymenoptera stings on a relative scale of '1' (low pain level) to '4' (extreme, excruciating, debilitating pain); a baseline was set by allocating a single honeybee sting a value of '2'. Schmidt sought out a variety of wasps, bees, and ants and used himself as a sting substrate, eventually ranking his responses to the stings of 96 species. He initially used a lower number of species by 1983, and his surveys led to a book (Schmidt 2016) expanding information on insect stings and venom. Considerable media attention arose from the pain scale, and the picturesque descriptions Schmidt gave—such as for the Red fire ant, *Solenopsis invicta* (level 1: 'Sharp, sudden, mildly alarming. Like walking across a shag carpet and reaching for the light switch') and, at the other extreme for the Bullet ant, *Paraponera clavata* (level 4: 'Pure, intense, brilliant pain. Like walking over flaming charcoal with a 3-inch nail embedded in your heel').

Nevertheless, direct economic or social return from insects gives them positive values and appreciation. Particularly when intermingled with cultural or traditional values these may become difficult to distinguish or allocate. Many insects traditionally eaten by people, for example are also used in traditional medicine. Increased or sustained demand for either purpose may lead to overexploitation and habitat destruction, or moves to increase supply through mass production. The balance of need can differ across taxa and regions, so that insects as medicine are strongly represented in Chinese culture, but far less so in parts of Europe. 'Entomotherapy' incurs similar responses to Entomophagy, despite an equally long history, and that some insects are used for both purposes—honeybee brood, royal jelly, and honey all have medical applications to some ethnic groups, but widespread social reluctance to use insects as medicine persists in Western society. Some medical applications of long standing are undergoing a resurgence. The use of fly maggots to debride dead wound tissues was used in the sixteenth century and, after increased use following the First World War declined as new antibiotics and surgical techniques were developed. Renewed use of 'maggot therapy', essentially a form of myiasis, reflects three aspects of wound healing: (1) debridement by dissolving necrotic tissue, with the maggots not eating healthy tissue; (2) wound disinfection by the maggots secreting antibacterial substances; and (3) accelerating healing by stimulating granular tissue production after the necrotic tissue has been eaten. A number of commercial companies produce 'medical grade maggots' bred from sterilised eggs, and for which demand is increasing.

An enormous variety of insects contribute to either entomophagy or entomotherapy, and the full extent of traditional uses of insects for food or medicine is still to be evaluated. A broad survey by Siddiqui et al. (2023) reported 235 medicinal insect species across 15 orders. Even more broadly, a major survey through the United Nations Food and Agriculture Organisation (FAO) estimated that insects form part of the traditional diets of at least 2 billion people, and that more than 1900 species have been reported as food (van Huis et al. 2013). The three most heavily consumed orders were Coleoptera (31%), larval Lepidoptera (18%), and Hymenoptera (14%). The greatest number of medicinal species in Siddiqui's account were Hymenoptera, followed by Coleoptera, Orthoptera, Lepidoptera, and Blattodea. In Nagaland (North-East India), seven ethnic groups used insects of 11 orders (and 28 families) as folk medicine in treatments of at least 50 human ailments (Mozhui et al. 2021), with the specific insects used varying with their availability in different places. The traditional knowledge is passed down verbally from generation to generation, but could be lost if not documented: Mozhui et al. considered that it has the potential to lead to development of novel drugs and treatments—a principle underlying much interest in insects as medicine. 'Prospecting' insects for beneficial pharmaceuticals is a commonly advanced reason for conserving insects, and numerous examples in ethnoentomological culture display the great variety of insects in traditional medicines that might support investigation. In China, with a strong tradition of folk medicine, the range of medicinal insects includes more than 300 insect species across 13 orders. Many of these are still underexplored for their pharmaceutical properties (Zhang et al. 2023), but Zhang et al. also noted that some of the insects with medicinal benefits are also known as pests needing suppression and control. They commented: 'turning these "agricultural/urban pests" into "medicinal treasures" has become a unique industry in China'. Such transformation in attitude has potential to become far more widespread.

Popularisation of insects is a complex but important aspect of promoting conservation interest. Rendering the alien world they represent to many people more familiar has many strands—not least by giving insects common names to replace the rather daunting Latin binomials used habitually by scientists! Common names can, indeed, be a means of enlisting local interest by basing them on places of concern (the Eltham copper butterfly is thus named for the Melbourne outer suburb where its conservation was first addressed, and this name is far more popular than *Paralucia pyrodiscus lucida*!), and attempts to provide common names for the more popular national insect groups (notably butterflies and dragonflies) have met with mixed success and adoptions. Thus, their adoption in North America was opposed by many entomologists who 'argued that some of the common names proposed were obscure, more complicated than the Latin names, and unlikely ever to be used by many' (Pearson 2013). Common names are an important inclusion for field guides intended for naturalist use, making them more 'user-friendly'. As one example, all 330 or so species of Australian dragonflies and damselflies are treated with both Latin and common names by Theischinger et al. (2021).

Fostering interest in insects among young people and educating them about insect significance is advocated widely, and the examples cited for Japan (Kawahara and Pyle 2013) included a range of insect-based video games, the substantial 'pet beetle' trade and numerous media presentations that emphasise the roles of insects in the nation's culture.

References

Bashir NH, Chen H, Munir S, Wang W, Chen H (et al.) (2022) Unraveling the role of lac insects in providing natural industrial products. Insects 2022, 13, 1117. https://doi.org/10.3390/insects13121117

Basset Y, Lamarre GPA (2019) Toward a world that values insects. Science 364(6447):1230–1231

Berenbaum M (2008) Insect conservation and the Entomological Society of America. Am Entomol 54:117–120

Berenbaum MR, Leskovsky RJ (2009) Chapter 174—movies, insects in. In: Resh VH, Carde RT (eds) Encyclopedia of insects, 2nd edn. Academic Press, Burlington MA, pp 668–674

Bradshaw CJA, Leroy B, Bellard C, Roiz D, Albert C (et al.) (2016) Massive yet grossly underestimated global costs of invasive insects. Nat Commun 7:12986. https://doi.org/10.1038/ncomms12986

Brock RE, Cini A, Sumner S (2021) Ecosystem services provided by aculeate wasps. Biol Rev 96:1645–1675. https://doi.org/10.1111/brv.12719

Cardoso P, Erwin TL, Borges PAV, New TR (2011) The seven impediments in invertebrate conservation and how to overcome them. Biol Conserv 144:2647–2655. https://doi.org/10.1016/j.biocon.2011.07.024

Carle E (1969) The very hungry caterpillar. Hamish Hamilton, London

Da-Silva ER, Coelho LBN, de Campos TRM, Carelli A, de Miranda GS (et al) (2014) Marvel and DC characters inspired by arachnids. Comic Grid J Comics Schol 4: 1–14. https://doi.org/10.5334/cg.aw

Donkersley P, Ashton L, Lamarre GPA, Segar S (2022) Global insect decline is the result of wilful political failure: a battle plan for entomology. Ecol Evol 12:e9417. https://doi.org/10.1002/ece3.9417

Duffus NE, Christie CR, Morimoto J (2021) Insect cultural services: how insects have changed our lives and how we can do better for them. Insects 12(5):377. https://doi.org/10.3390/insects12050377

Fukano Y, Soga M (2021) Why do so many modern people hate insects? The urbanization-disgust hypothesis. Sci Tot Environ 777:146229. https://doi.org/10.1016/j.scitoitenv.2021.146229

Golden DBK (2015) Anaphylaxis to insect stings. Immun All Clin North Am 35:287–302. https://doi.org/10.1016/j.iac.2015.01.007

Guiney MS, Oberhauser KS (2008) Insects as flagship conservation species. Terr Arthropod Rev 1:111–123

Hall DM, Martins DJ (2020) Human dimensions of insect pollinator conservation. Curr Opin Insect Sci 38:107–114

Kawahara A, Pyle RM (2013) An appreciation for the natural world through collecting, owning and observing insects. In: Lemelin RH (ed) The management of insects in recreation and tourism. Cambridge University Press, Cambridge, pp 138–152

Kellert SR (1993) Values and perceptions of invertebrates. Conserv Biol 7:845–855. https://doi.org/10.1046/j.1523-1739.1993.740845.x

Leather SR (2015) Influential entomology: a short review of the scientific, societal, economic and educational services provided by entomology. Ecol Entomol 40:36–44. https://doi.org/10.1111/een.12207

Lockwood JA (2013) The infested mind. Why humans fear, loathe and love insects. Oxford University Press, New York

Losey JE, Chen C, Davis AE, Deitsch JF, Gertin JG, Gorneau JA (et al.) (2022) Insects and spiders on the web: monitoring and mitigating online exploitation of species and services. Glob Ecol Conserv 36, e02098. https://doi.org/10.1016/j.gecco.2022.e02098

Mozhui L, Kakati LN, Meyer-Rochow VB (2021) Entomotherapy: a study of medicinal insects of seven ethnic groups in Nagaland, North-East India. J Ethnobiol Ethnomed 17(17):1–22. https://doi.org/10.1186/s13002-021-00444-1

Nazari V (2021) Taxonomy at face value: an assessment of entomological postage stamps as effective teaching aids for science educators. Res Idea Outcomes 7:e68056. https://doi.org/10.3897/rio.7.e68056

Pearson DL (2013) Tiger beetles: lessons in natural history, conservation and the rise of amateur involvement. In: Lemelin RH (ed) The management of insects in recreation and tourism. Cambridge University Press, Cambridge, pp 56–75

Samways MJ (2018) Insect conservation for the twenty-first century. In: Shah MM, Sharif U (eds) Insect science—diversity, conservation and nutrition. InTech, pp 19–40. https://doi.org/10.5772/intechopen.73864

Samways MJ, Lockwood JA (1998) Orthoptera conservation: pests and paradoxes. J Insect Conserv 2:143–149

Schmack JM, Egerer M, Karlebowski S, Neumann AE, Sturm U (2024) Overlooked and misunderstood: how urban community gardeners perceive social wasps and their ecosystem functions. J Insect Conserv 28:283–289. https://doi.org/10.1007/s10841-024-00548-5

Schmidt JO (2016) The sting of the wild. Johns Hopkins University Press, Baltimore

Siddiqui SA, Li C, Aidoo OF, Fernando I, Haddad MA (et al.) (2023) Unravelling the potential of insects for medicinal purposes—a comprehensive review. Heliyon 29: e15938. https://doi.org/10.1016/j.heliyon.2013.e15938

Sumner S, Law G, Cini A (2018) Why we love bees and hate wasps. Ecol Entomol 42:836–845. https://doi.org/10.1111/een.12676

Theischinger G, Hawking J, Orr A (2021) The complete field guide to dragonflies of Australia, 2nd edn. CSIRO Publishing, Melbourne

van Huis A, van Itterbeeck J, Klunder H, Mertens E, Halloran A (et al.) (2013) Edible insects: future prospects for food and feed security. FAO Forestry Paper 171, Rome

Vanderstock A, Grandi-Nagashiro C, Kudo G, Latty T, Nakamura S (et al.) (2022) For the love of insects; gardening grows positive emotions (biophilia) towards invertebrates. J Insect Conserv 26:751–762. https://doi.org/10.1007/s10841-022-00419-x

Wilson EO (1987) The little things that run the world (the importance and conservation of invertebrates). Conserv Biol 1:344–346

Yen AL (2009a) Edible insects: traditional knowledge or western phobia? Entomol Res 39:289–298. https://doi.org/10.1111/j.1748-5967.2009.00239.x

Yen AL (2009b) Entomophagy and insect cnservation: some thoughts for digestion. J Insect Conserv 13:667–670. https://doi.org/10.1007/s10841-008-9208-8

Zhang E, Ji X, Ouyang F, Lei Y, Deng S (et al.) (2023) A mini review of the medicinal and edible insects from the traditional Chinese medicine (TCM). Front Pharmacol 14: 1125600. https://doi.org/10.3389/fpharm.2023.1125600

Chapter 6
The Reasons for Insect Conservation

6.1 Introduction

The importance of insect conservation is often difficult to justify to a hostile or non-sympathetic audience! Summarising the importance of insects, and thus establishing the rationale for their conservation, in the wider community is difficult but necessary. Collectively, the services provided by 'entomology' embrace a great diversity of themes—Leather (2015) noted major categories as 'scientific', 'economic', 'societal', and 'educational', for example. Earlier chapters have emphasised several aspects of insect 'importance', and justifying the need for conserving insects to an often uninformed (and cynical!) audience may draw on a number of those complementary themes. They range from philosophical to wholly pragmatic, are not mutually exclusive, and interweave with many facets of the wellbeing of humanity and the natural world. Simply 'liking' an insect (Chap. 5) is sufficient to want to conserve it, but is essentially a subjective, perhaps individualistic, reason that may be difficult to disseminate widely. The major themes that justify insect conservation more widely are discussed in this chapter. The values of insects are increased massively by the different life stages—for some, the larval stage is the most tangibly appreciated—for others, adults have value, and for many taxa both or all life stages contribute to economy or wellbeing.

The core importance of insects to humanity and to Earth's terrestrial and freshwater ecosystems necessarily underpins the wisdom of informed conservation, but with primary emphasis differing widely across different sectors of society. The range of current concerns, whether loss of taxa or ecosystems, non-target impacts of pest management, cultural or economic loss, or others, is complemented by the less tangible implications that in the future insects may continue to yield other benefits such as 'new' drugs, antimicrobials, or medicines that would otherwise never be known, and additional nutritional benefits, as discussed earlier. Whereas many of the reasons for insect conservation devolve on insects as the 'targets' for

T. R. New, *Insect Conservation in Australia: Why and How*,
https://doi.org/10.1007/978-3-031-66631-5_6

conservation, the wider significance of many insects as symbols of wider environmental conditions also gives them important roles such as 'indicators' valuable for assessing and monitoring environmental health. They thus become 'tools' in a far wider conservation endeavour, and important as monitors of environmental condition, its changes from human activity, and the process of recovery as conservation measures take hold.

Establishing an optimal and persuasive rationale and process for insect conservation is thus a complex and somewhat subjective exercise. But, in discussing the need to 'future-proof' insect life, Samways (2020) noted six inter-related 'anchors' for conservation which he summarised as (1) philosophy (how we value nature and our relationships with it); (2) research (discovering what is needed for strategic conservation action); (3) psychology (focusing on people's perceptions, aspirations, and shared goals, and how insect conservation 'fits'); (4) policy (translating research into viable and affordable options for practical conservation); (5) practice (the actual activities of conservation); and (6) validation and monitoring (assessing progress and, wherever necessary, adjusting goals and actions to provide the best outcomes). All depend on acceptance that insects are worth conserving, the main theme of this chapter.

6.2 Philosophy

Two contrasting philosophies affect how people consider or perceive insects. Utilitarian (or instrumental) values of insects to humanity imply that 'nature' is for humans to exploit and use, either directly or to satisfy needs (such as cultural, aesthetic, scientific) without diminishing it directly. Intrinsic value acknowledges that every life form has worth, irrespective of its value to humans. In essence, philosophy—how we value insect life—is the ethical foundation for insect conservation.

Initially, and sometimes not persuasive to pragmatic recipients, is the idealistic maxim that no species should be allowed to become extinct from human cupidity, so that there is an ethical duty to ensure that this does not occur. Many people, whilst willing to broadly accept the principle, make exceptions for insects and other invertebrates—and many entomologists pass their careers in efforts to suppress or eradicate economic or other pest insect species that may interfere with human food or commodity production, health, or standard of living. Proposals to conserve pest species or disease vectors can have limited traction. Indeed, harmonising insect conservation (broadly, 'saving insects') and pest management ('killing insects') can induce many conflicts (Kogan and Lattin 1993). Conservation typically focuses on native insect species, but some native insects can become serious pests of high-value crops and create economic loss; they may then be targeted for local or wider suppression with possible local extinction. In such contexts, ethical and moral considerations become secondary to economic issues and protection of human food supply. The perception can differ with alien insect pests, for which only poor grounds for conservation in areas they have invaded are persuasive, and the major purpose is to eradicate them or, more practically, to reduce their influence to below any economically damaging threshold level. For any insect pest management programme, it may be desirable to conserve small pest populations as reservoirs for natural enemies (even when these have been introduced) that can then disperse to attack the pest on crops when its numbers increase, as a facet of conserving natural enemies. Integrated pest management will often seek to increase impacts of both native and any introduced natural enemies on the target pest and provide for their wellbeing over periods of pest absence. Increasing native vegetation diversity and structure in field surrounds to provide a supply of alternative native prey/hosts and flowers with nectar/pollen for natural enemies are well-tested conservation steps in agricultural landscapes. Increasing diversity of pest management strategies—using biological control by introduced agents or an increasingly sophisticated array of microbial diseases—is likely always to suggest a possibility of non-target impacts and their use may lead to an 'uneasy truce' between practitioners and conservationists. As noted in Chap. 1, the large number of insects killed as bycatch by sampling, even during conservation exercises, does little to support advocacy for conserving insect diversity.

6.3 Culture

Insects have been intertwined with human art and culture since before the start of recorded history, their roles ranging from fundamental beliefs, religion, mythology and folklore, and art to derived appreciations of practical uses as commodities and in trade. References to bees occur across most ancient human cultures, indicating skilled beekeeping approaches, and the many other insects noted imply widespread early appreciation of insect variety and significance. That variety, of both contexts and insect taxa, extends across the entire human awareness of the natural world (Kritsky and Smith 2018).

Building on cultural and traditional values of insects can help diversify their roles in modern society. Expansions of entomophagy and entomotherapy, for example, draw heavily on species used for many centuries and are part of 'folk culture' but also progressively absorb further taxa as their use expands. Measures to safeguard individual species that are locally significant to people are likely to be sympathetically supported. The high cultural standing of dragonflies in Japan has been instrumental in their acceptance as worthy of conservation and the establishment of aquatic systems and tourism attractions based on viewing these insects. 'Iconic' insect species or groups may be adopted as flagships for conservation, drawing on their more widely accepted inherent 'worth'. Measures to conserve them may benefit numerous other species.

Long traditions of insects in national or regional 'culture' take many forms; Chinese jade carvings of butterflies, many as pendants or buckles, have romantic symbolisms, whilst jade cicadas were used as funerary objects, placed in the mouth of deceased people and associated with rebirth and immortality. Reflecting this, jade cicada pendants are symbols of purity and integrity. Luminescent fireflies (Coleoptera: Lampyridae) are celebrated widely in art, for watching as a recreational activity and a tourism focus (p. 47). Fireflies are an active global conservation focus, with activities coordinated through an IUCN Specialist Group and considerable progress over the last two decades (Lewis et al. 2024).

Cicadas and crickets are valued for 'sporting' activities such as singing contests (cicadas and crickets) and fighting (crickets). The last activity accelerated in China during the Tang Dynasty (618–906 AD) and, whilst gambling is illegal, organised series of cricket fights remain widespread and popular; one press article reported that more than $60 million was spent on the cricket hobby in China in 2010, for example. Such values lead to intergrading of cultural and economic worth.

6.4 Economy

The economic benefits from insects range from local exploitation and use to wider national and international production, trade, and exchanges. Overlapping with 'culture' above, use of insects as personal ornaments or jewellery can transform into exploitative trade with brightly coloured beetles and butterflies, in particular, sought for sale. 'Butterfly wing pictures' have been popular tourist items across much of the tropics, for example. Conservation concerns embrace many of the themes discussed earlier, some overlapping with moves to increase the popularity of insects.

6.4.1 Insects in Ecotourism

Growth of international ecotourism, with strong local financial benefits to tour operators, travel guides, hotel and restaurant proprietors, and souvenir sellers, is commonly associated with wishing to see spectacular vertebrates—the 'big five' in southern Africa is a classic focus—but some insect examples, largely 'butterfly-watching', extend the more confined experiences of butterfly houses to the wild. Butterfly tours occur in many countries throughout the tropical zones, mostly advertised on grounds of seeing diversity and viewing many species.

Others are more strongly focused. The tourist industry on Rhodes, Greece, is supported strongly by a reserve known as 'Valley of the Butterflies', where tourists can view massed aggregations of the colourful Jersey tiger moth (*Euplagia quadripunctaria*, Erebidae, as an 'honorary butterfly'). However, there, as elsewhere, disturbance by visitors can be harmful, and people are asked not to clap, whistle, or otherwise make loud noises that cause the moths to fly. Likewise, many tourists in Mexico visit the traditional overwintering sites of the Monarch butterfly (*Danaus plexippus*, Nymphalidae, in Australia termed the Wanderer), and several tourist agencies take only small groups of people at any time, and emphasise the need for quiet movements in the major reserve, the 'Reserva Mariposa Monarca'. Some

Mexican sites are under continuing threat from logging, and tours provide the opportunity to encourage interest in conservation, advocating or practising site protection and restoration through tree planting; one press report noted 160,000 visitors in 2021. More generally, because many such insects with tourist appeal are very seasonal in appearance, tours may operate for only short parts of the year. For some taxa and sites, firm control on access may be needed to reduce disturbance.

Some less conspicuous insects are also tourist foci. Crepuscular fireflies are the prime target of tours in Indonesia, Philippines, Malaysia, and other countries in south-east Asia, some advertised as part of a river cruise or longer day tours. Related fireflies or cave glowworms (Diptera: Keroplatidae, p. 47) attract tourists in North America, Australia, New Zealand, and elsewhere. The most visited southern hemisphere glowworm site (Waitomo Glowworm Cave, on the North Island of New Zealand, harbouring *Arachnocampa luminosa*) attracts more than 500,000 visitors a year (cited by Merritt and Clarke 2013), and major commercial glowworm-viewing guided tourist operations exist in Tasmania (*Arachnocampa tasmaniensis*) and Queensland (*A. flava*), one operation in the latter state involving an artificial constructed cave environment. In Australia, glowworm tourist activities have been cited as 'a multimillion-dollar industry', with need to protect the cave environments and regulate visitor behaviour and numbers (Hall 2013). The glowworms have been described as 'charismatic microfauna', and the species are all narrow range endemics, some known from only single wet cave systems and occurring also in rainforest gullies along the east coast of Australia.

Insects are increasingly considered in ecotourism, in the varied contexts of (1) focus on a single species or phenomenon; (2) paired with another ecotour focus; or (3) primarily research-based as 'information gathering' from observations.

6.4.2 Insects in Trade

The considerable take of common insects for the tourist trade is difficult to monitor and regulate, and its impacts are largely overlooked. It is likely that the great majority of these are captured directly rather than reared. Far greater attention has been paid to the possible overexploitation of commercially desirable collectable insects such as rare butterflies and beetles, which are advertised at prices up to several thousand dollars for particularly desirable specimens (particularly of the rarest birdwing butterflies and largest stag beetles), and can involve direct destruction of habitat to acquire them.

'Listed' species for which collecting is prohibited can also attract high prices on the 'black market'—not least because prohibition of capture can increase desirability and prices, and small size renders dead insects 'easy' to smuggle. Rather few insects are listed on CITES (the Convention on International Trade in Endangered Species) for trade restriction, but amongst these, rarer birdwing butterflies, in particular, are highly sought by collectors, and their values—as with other rare butterflies—attract illegal trade which is difficult to monitor (Collins and Morris 1985). In the early 1990s the global butterfly trade was estimated at about US$ 100 million/year (Parsons 1992). Attempts have been made to centralise trade (so that it could be monitored and legalised) and to use 'ranched' butterflies for this (to avoid continued plunder of wild populations), most notably through the formerly pioneering Insect Farming and Trading Agency (IFTA) founded in 1978 in Papua New Guinea. The principle involves individual people who plant larval food plants that attract female birdwings to oviposit, rear the caterpillars in cages (so avoiding the natural mortality from predators and parasitoids), and are being paid for a proportion of the reared adults as pristine specimens for sale to collectors, whilst releasing the remainder back into the natural forest habitat. As Parsons (1992) discussed, these systems can provide remote rural communities with additional income, but also promote forest conservation through reducing human pressures. The wide-ranging implications of IFTA implicitly included conservation (Table 6.1). Such 'umbrella organisations' have been emulated elsewhere, but have proved difficult to sustain (Parsons 1992; New 1997). Without some controls on direct collecting of rare commercially desirable insects, trade may add to the already existing threats from unregulated habitat losses; with controls such as the above, local people may benefit, and the local environment be protected more effectively as a recognised source of income.

Monitoring insect trade, even with formal systems involving CITES species such as the above, is difficult, as Peggie (2022) showed for Wallace's birdwing (*Ornithoptera croesus*) in Indonesia. Records of numbers exported to various countries as reported in the CITES database from 2011 to 2017 and numbers imported

Table 6.1 The reasons and objectives for foundation of the Insect Farming and Trading Agency (IFTA) in Papua New Guinea (PNG) (Parsons 1992)

1.	To promote, through extension trips, the protection of butterflies and other insects in less advanced areas of PNG as an alternative source of income for a large number of subsistence farmers (in line with the Government's decentralisation policy)
2.	To restrict the insect trading business to citizens only
3.	To provide basic collecting and storage equipment to collectors/farmers
4.	To ensure that fixed and reasonable prices are paid to collectors/farmers
5.	To ensure that payments are made more expediently than they would be if the collectors/farmers had to wait for monies to be returned via overseas mail (losses of mailed cheques also being known)
6.	To provide a centralised body within PNG with which both citizens and buyers can directly and more easily communicate than if they had to do so with each other
7.	To act as an official agent in business dealings with overseas buyers in order to better deter payment defaults by buyers
8.	To ensure the best quality of incoming (and outgoing) stock, including the addition of locality data to each specimen
9.	To pool stock in order to supply the larger orders of dealers
10.	To assume some control over demand by the manipulation of outgoing stock
11.	To act as an educational centre for the instruction of citizens in insect farming and trading methods
12.	To ensure that insects are treated as a renewable resource
13.	To promote conservation of butterflies and their habitats

by those countries were in most case not the same. Thus, for *O. croesus* exported from Indonesia to Australia, numbers reported by Peggie included (export report number, import report number): 290, 42 (2011), 56, 202 (2012), 67,190 (2013), and 0, 290 (2014)—such figures illustrate the difficulties of precise trade assessments.

6.4.3 Insects in Medicine

The allure of financially lucrative 'new drugs' or other health benefits from insects is a foundation for 'insect biotechnology'. Known also as 'yellow biotechnology' (named after the colour of insect haemolymph), this can be defined more broadly as 'the use of whole insects, their organs, cells or molecules, but also their symbiont microbes, in the fields of medicine ('red biotechnology'), agriculture ('green biotechnology') and industry ('white technology')' (Hoffmann 2017). Advances have been driven largely by developments in molecular biology. Interests are founded in the massive array of active compounds that insects have developed and that defend them against diseases or enemies or that influence their food sources and selection. Simplistically, the vast breadth of foodstuffs that insects collectively exploit is associated with a correspondingly large array of enzymes, with the possibility that every insect species is in some way 'chemically unique', in its complement of biologically active natural products. These are synthesised by either the insect itself or its associated microorganisms, or derived from plants. The hope that these natural substances may be useful to humankind in fields such as drug discovery and plant protection has given their exploration massive economic potential and interest. Potential rewards from such bioprospecting are a powerful support for conserving insect diversity for possible future, and as yet unanticipated, needs. As Dettner (2011) put it 'Worldwide … arthropods therefore appear to represent an unexplored and unexploited, nearly inexhaustible source of drugs for modern medicine'. The many contexts of interest reflect an insect's lifestyle and feeding habits. As examples noted by Dettner, (1) blood-feeders (such as some bugs, lice, mosquitos, a range of biting flies) must keep the host blood in liquid form, so excrete anticoagulants in their saliva; (2) the maggots of the fly *Lucilia sericata* clean wounds both mechanically and chemically—by 'biosurgery'—and may produce antimicrobial peptides; (3) as an example of thriving in an 'extreme environment', larvae of petrol flies (*Helaeomyia*) live in petroleum pools, and may have some bacterial strains and enzymes that enable tolerance to some organic compounds; and (4) some biologically active insect peptides are antimicrobial or antiviral. Indeed, 'The potential pharmaceutical and industry applications of the symbiotic microbial community in extremophile Diptera are better explored than are those of dipteran chemistry' (Adler and Courtney 2019).

Whilst critics have suggested that advocacy for insect pharmaceutical values is exaggerated in conservation, the above and other examples indicate the wisdom of continued investigations—and of assuring that the insect target species remain alive to be investigated.

6.4.4 *Insects as Food and Feed*

Cultural and religious traditions have incorporated insects as human food throughout much of the world. It has been estimated that insects are included in the traditional diets of at least 2 billion people, with more than 1900 species reported as foods (van Huis et al. 2013) and spanning a wide range of orders. That major report and overview undertaken through FAO was a major impetus for entomophagy, introducing the three major reasons for promoting insects as food and feed (Table 6.2) and emphasising insects as an abundant natural resource amenable to mass-rearing operations that could contribute to alleviating future food insecurity—not least through their high feed conversion efficiency. This approach could also dramatically reduce the amounts of waste material from other sources—feeding on waste and converting it into nutritious food or feed ('upcycling') may become important in fields such as urban waste management. Production and quality control of insect mass-rearing operations are highly regulated, with public health and food safety both major concerns.

A major advance occurred with the European Union's approval (in 2017) of insect protein as an ingredient in aquafeed, stimulating capital investments in insect mass production systems: Ndondo (2022) quoted this investment as near Eu1 billion in 2020. Continually innovative production systems are likely to increase large-scale insect farming as societal needs for meat and protein products grow to a level

Table 6.2 The three major reasons for promoting entomophagy (van Huis et al. 2013)

Health	Insects are healthy, nutritious alternatives to mainstream staples such as chicken, pork, beef, and even fish (from ocean catch) Many insects are rich in protein and good fats and high in calcium, iron, and zinc Insects already form a traditional part of many regional and national diets
Environmental	Insects promoted as food emit considerably fewer greenhouse gases than most livestock (methane, for instance, is produced by only a few insect groups, such as termites and cockroaches) Insect rearing is not necessarily a land-based activity and does not require landclearing to expend production Feed is the major requirement for land The ammonia emissions associated with insect rearing are also far lower than those linked to conventional livestock, such as pigs Because they are cold-blooded, insects are very efficient at converting feed into protein (crickets, for example, need 12 times less feed than cattle, four times less feed than sheep, and half as much feed as pigs and broiler chickens to produce the same amount of protein) Insects can be fed on organic waste streams
Livelihoods (economic and social factors)	Insect harvesting/rearing is a low-tech, low-capital investment option that offers entry even to the poorest sections of society, such as women and the landless Minilivestock offer livelihood opportunities for both urban and rural people Insect rearing can be low-tech or very sophisticated, depending on the level of investment

that demands cannot be met from conventional but environmentally expensive food production systems, and viable alternatives are sought.

Much of this initial endeavour has focused on a few amenable species from which benefits are sufficiently evident to attract investment. One important example is the Black soldier fly (*Hermetia illucens*, Stratiomyidae), the first species adopted for a large-scale commercial insect farming project in the United Kingdom. Probably originally a Neotropical species, the fly is now very widespread. Females oviposit on decaying organic material; eggs hatch within about 4 days, and the five larval instars are completed in about 12–18 days under good conditions but can take much longer. Applications and values of *Hermetia* are numerous (Table 6.3) and continue to proliferate (Kaczor et al. 2023), but perhaps the key interest is the reductions in organic waste such as compost by around half to produce a valuable feed (as larvae and pupae) for domestic animals, as 'bioconverters'. The acceptable 'wastes' for larval food include dead animals, plants, kitchen and slaughterhouse materials, municipal waste, reflecting the ability of fly larvae to digest a variety of materials. An additional benefit is that the fly is not a vector of any disease or parasite.

Many parallels occur in other insects potentially or actually valuable in the food/feed industries, and their number continues to expand as additional candidates are explored for nutritional content and amenability to economical mass-rearing. Mealworms (Coleoptera: Tenebrionidae) are one such case, with additional species beyond the common mealworm (*Tenebrio molitor*, reared as commonly as *Hermetia*) receiving critical appraisal. The 'Superworm' (*Zophobas morio*) is a far larger mealworm, with larvae up to about 55 mm long, and has already proved valuable in fish food and for poultry. Rumbos and Athanassiou (2021) suggested that *Zophobas* is a very promising insect protein source with its potential yet to be realised.

A key consideration for any insect that is to be mass-reared—for food or any other purpose—is that it must develop rapidly to reduce the overhead costs of long-term rearing without return. Much of the foundation of rearing techniques for

Table 6.3 The actual and potential values and uses of Black soldier fly, *Hermetia illucens*, illustrating benefits of mass-rearing operations (after Kaczor et al. 2023)

Bioconversion of organic wastes	Larval feeding on a variety of wastes including food waste and sewage sludge
Animal feed	Wide variety of contexts; mainly using larvae and pupae, whole or ground as feed supplements
Chitin/chitosan source	Wide range of industrial uses
Antimicrobial properties	Decontamination of faeces, sewage waste
Biodiesel production	From feeding on organic biomass: Reuse of fats
Biogas production	From rearing wastes and residue
Entomoremediation	Detoxify pollutants from soil, sediments, and organic biomas
Insect frass	Use as fertiliser and for antifungal properties
Larvae as food	For people
Cosmetics/personal care products	Chemical derivatives
Bioplastics	Biodegradable plastics from soldier flies

entomophagy has drawn on the long-term experiences with mass-rearing insect biological control agents for which similar 'factory' objectives of quantity, safety, and quality assurance apply.

6.4.5 Additional Benefits

A number of more 'unconventional' ecosystem services have been noted for insects and add to the potential of insects to benefit humanity and the world's environments. Two early-stage examples discussed by Morimoto (2020) are (1) the degrading of plastic by larvae of two Wax moth species (*Plodia interpunctella*, *Galleria mellonella*, Pyralidae), and of styrofoam and polystyrene by mealworm larvae (*Tenebrio molitor*, Tenebrionidae), suggesting the potential to contribute to control of plastics pollution, and (2) the high lipid content of fat that can be extracted from some holometabolous insect larvae has been suggested to make those larvae a suitable source of 'biofuel' as renewable energy. Morimoto noted that larvae could be fed on manure and other organic material (as for *Hermetia*, above).

The full range of potential applications of insect life to human welfare is impossible to assess—but the diverse themes suggested or developed across a variety of fields until now emphasise the importance and wisdom of a 'pragmatic insurance' approach to prevent insect losses. Some of these themes draw on unusual biological features, such as relevance of insects in cryobiology. Leather (2015) also noted that insects have inspired engineers, leading to developments such as microdrones (from dragonflies and bees), airless tyres (from honeycomb design), and water collection in desert regimes (from the adaptations of South African desert beetles that can collect droplets of fog on their elytra).

The use of bee hives to manipulate elephant behaviour illustrates an innovative benefit of insects to people. In Tanzania, establishment of bee hives along the boundary between the Udzungwa Mountains National Park and adjacent cropping areas helped to reduce crop damage by elephants by deterring them from moving into the crops (Scheijen et al. 2019). Suspended hives were deployed 10 m apart and linked by wire; shaking of the wire by elephants stimulated bee activity. Local farmers benefited from both crop protection and income from the hives.

6.5 Ecology

The varied ecological interactions in which insects participate and the 'services' these provide continue to be documented and evaluated. The consequences of diminishing these is perhaps the most politically and practically persuasive reason for modern insect conservation, with a range of strong economic and welfare strands, and benefits to humanity a major complementary driver of interests.

The cash values of ecological services are difficult to estimate, not least because of fluctuating market prices for agricultural commodities. Estimates have been made for several contexts, but any comprehensive evaluations are elusive; thus, the European Commission (2023) commented that 'most of the essential benefits that pollinators provide remain unquantified' but still ventured that animal pollination contributed at least EUR 5 billion/year to the EU's agricultural output. Losey and Vaughan (2006) defined 'value' by 'documented financial transactions – mostly the purchase of goods or services—that rely on these insect-mediated services', and did not include services from domesticated species (such as honeybees) or mass-reared biological control agents in their pioneering calculations. They provided value estimates for four major services selected because sufficient data were available (dung burial, pest control, pollination, and wildlife nutrition) to derive an annual figure for the United States of 'at least US$57 billion'. This figure has been quoted extensively as a 'wake-up call' for appreciating insect values and the need to conserve these. Despite a substantial lack of data on many regionally important crops assuring food security, especially in developing countries, and other uncertainties more recently estimated global values of crop pollination services (adjusted for inflation in March 2020) were far higher, ranging from US$ 195 billion to around US$ 387 (267–675) billion a year (Porto et al. 2020). The related topic of evaluating costs of the impacts of invasive insects, many of which are 'pests', emphasises the difficulties of obtaining 'hard' figures for such broad contexts: Bradshaw et al. (2016) noted a minimum of US$ 70 billion annually but suggested that this was likely to be 'gross underestimation'.

The various caveats and uncertainties discussed by Losey and Vaughan illustrate the difficulties of gaining precise values for any of these services. As examples, (1) for pest control, data were restricted to suppression of insect pests that attack crop plants by insect predators and parasitoids, and omitted the data-poor impacts of insects suppressing plant weeds and reducing populations of insects that attack livestock and humans, and (2) 'wildlife nutrition', relating to activities allied to hunting, fishing, and observing wildlife, involving numerous species of insectivorous birds, fish, and small mammals, with the major service of concentrating and moving nutrients through food webs in many different habitats. This last was inferred to be the predominant component of insect ecological services, with freshwater recreational fishing constituting nearly 28 billion dollars of the total, with Americans then spending an estimated US$49 billion annually on this activity portfolio.

Manipulations of services from enhancing crop pollination (such as by migratory beekeeping—the seasonal shifting of hives to track pollination needs)—and increasing pest control by native or specifically introduced natural enemies demonstrate well the impacts that particular insects may have in anthropogenic ecosystems, and the practical benefits of sustaining them. They add enormously to the financial benefits suggested by Losey and Vaughan. Conservation of ecological services providers has rather different emphasis from conserving endangered insect species (Losey and Vaughan 2008) (Table 6.4), reflecting that most of the important service providers are still quite abundant—there are concerning exceptions such as losses of key pollinators—and that their functional roles are often performed by many species.

Table 6.4 The contrasted features of insect species that are 'endangered' (and the most frequent targets for conservation) and 'service providers', indicating that the conservation strategies may differ considerably (from Losey and Vaughan 2008)

	Endangered species	Service species
Focal unit	Single species	Species complex
Density	Very rare	Common
Origin	Native	Mixed
Economic impact	Variable	Positive
Result of inaction	Extinction	Lost resources

Even confining discussions of insect importance to the politically and publicly acknowledged foci of pollination, decomposition and nutrient recycling, and pest management by predators and parasitoids provides a compelling suite of reasons for insect conservation and greater understanding of their ecological roles. In short, the world as we know it could not continue without insect activities. Their declines from food webs potentially affect vast numbers of animals and plants that we do appreciate—as Goulson (quoted by Milman 2022) commented 'People might not give a damn about insects but they like pretty birds in their garden': without insect foods they also will be lost. Declines of insect prey are amongst the possible causes of woodland bird declines in south-eastern Australia, and Braby et al. (2023) have urged greater cooperation between entomologists and ornithologists to understand this scenario.

6.6 Insects as Conservation Tools

Insects that respond to environmental changes in some predictable way may be used as 'indicators', valuable in characterising environmental condition and monitoring changes. Some may be likened to the 'canary in the coalmine', through which early warnings of undesirable changes occur through changed richness, abundance, or incidence, and might apply far more widely than to the taxa detected. Insects can thus be used to assess environmental conditions and, through the repeated interval sampling of 'monitoring', to detect and assess changes in response to disturbances or conservation management activities. These roles give insects values in conservation activities extending widely beyond the insects themselves.

'Indicators' may be single species or larger groups, and the principle extends across all insect habitats. Ideally, indicator taxa should be well-known taxonomically (so, easy to identify), detect and sample, by easily replicable standard methods, be present over the areas to be assessed, and their responses to a given stressor understood and believed to reflect trends in other taxa. They thereby become measurable surrogates for wider taxa in any individual survey, with the potential to be used for comparative assessments across sites and wider areas, and allow for cost-effective, informative, and rapid surveys. As with any insect surveys, facility to avoid the needs for species-level identifications without loss of information and use higher taxonomic levels can be a considerable logistic advantage. Any developmental stage may be used when appropriate—for Odonata, for example, adults, larvae (usually termed nymphs in exopterygote insects), and the cast final instar larval moult (exuviae) left grasping emergent vegetation after adult emergence are all amenable to counting and standard sampling, and may be used singly or to complement each other, and can collectively increase the season over which data can be obtained.

In both lotic and lentic water bodies, much environmental assessment has been based on incidence and relative abundance of various groups of benthic invertebrates. Most usefully, especially for running waters, larvae of three widespread, diverse, and abundant insect orders have proved very useful reflections of water quality as the 'EPT index', for Ephemeroptera, Plecoptera, and Trichoptera, all groups that are widely sensitive to pollution, and combining the numbers of species (or morphospecies) as a proportion of the total taxa found. The indicator is based on the premise that high-quality waterbodies usually have the greatest richness of these taxa and that the score declines when pollution increases. In Victoria, stream condition can be evaluated by the broader 'Aquatic life sub-index' (DIPE 2014) that combines four aspects of macroinvertebrate incidence from standard samples from riffle and edge habitats but identified only to family level, from reference (high quality) and focal streams. This approach thus combines the four scores from the simple measures summarised in Table 6.5, but application depends on good reference

Table 6.5 An example of the way in which insects can be used to assess environmental quality: the Index of Stream Condition Aquatic Life Sub-Index developed for Victoria: summary from DIPE (2014). The Index is based on aquatic macroinvertebrates and has four indicator categories, as below

Category	Explanation
AUSRIVAS (Australian River Assessment System)	Consists of models that predict what macroinvertebrates should be present in specific stream habitats under reference condition By comparing the families predicted to occur with the number actually present, the observed/expected (O/E) index, which ranges from '0' (no expected families found) to '1' (all expected families found), can be calculated 'Key families' can be used as an alternative to AUSRIVAS, representing the families expected to occur in streams of good condition
SIGNAL (Stream Invertebrates Grade Number Average)	Families awarded a 'sensitivity score' according to tolerance/sensitivity to various pollutants, and SIGNAL score calculated by totalling scores for each family found and dividing by the number of families used; result helps assess site status in terms of pollution
EPT	Measure of Ephemeroptera, Plecoptera, and Trichoptera as three abundant sensitive orders; high-quality streams have highest richness, and this declines with pollution
Number of families	The number of families of macroinvertebrates at a site can be a reasonable representation of ecological health

The aquatic life sub-index score (out of 10) is calculated by the following, adding the four indicator scores:

Aquatic life sub-index = 10/16 (AUSRIVAS [or key families score] + SIGNAL score + EPT score + number of families score)

source documentation—that reference condition has been defined in Victoria, so that departures from it in the main bioregions can also be defined. 'Key families', for example, are those expected to occur in streams in good condition, with their absence likely to indicate lessening quality. For the many regions in which good baseline reference data are not available readily, the EPT index is used widely, as a 'stand-alone' measure. This can be augmented by counts of Odonata, as used by Kietska et al. (2019) in South Africa to also explore the related concept of 'umbrella species' and the consequent 'umbrella index', by which conservation of selected surrogate species in biodiversity assessments benefits co-occurring taxa.

An early definition of an umbrella species for conservation is 'a suitable target species whose habitat requirements can cover other species, thus the protection of the target species benefits co-occurring species' (Wilcox 1984), so that selecting a reliable set of these umbrella species might improve conservation efficiency. Most emphasis on selecting umbrella taxa has involved mammals or birds in terrestrial ecosystems, but aquatic macroinvertebrates have been examined in several places—such as China (Gao et al. 2020), North America, and Europe—and reflecting their diversity, ease of sampling, and reasonably good taxonomic understanding. In terrestrial ecosystems, butterflies have sometimes been proposed as umbrellas, most notably in restricted habitats such as remnant vegetation patches that are likely to support many other undetected or poorly known invertebrates. However, as Fleishman et al. (2000) observed, rare butterfly species may be poor umbrellas because their distributions are highly restricted and often cannot be influenced by managers. More generally, survey of some subset of a total fauna may serve as an umbrella for a wider community. They can then be 'an attractive shortcut' (Betrus et al. 2005) in land-use decisions, but validation may be needed before uncritical use.

The wider applications of terrestrial insects as 'indicators' draw on their varied responses and sensitivities, and applications vary greatly (Chowdhury et al. 2023). Three major categories—environmental indicators, ecological indicators, and biodiversity indicators—distinguished by McGeoch (1998) to reflect the predominant uses of bioindicators illustrate their variety. Perhaps the most complex of these is 'environmental indicators' for which McGeoch followed earlier workers in recognising five categories of responses to environmental change. The categories and the suitability, applications, and relevance of particular insect taxa have been debated extensively in a flurry of later papers and some leading to claims of values that were not strongly supported. Their major functions are summarised in Table 6.6, which also emphasises that indicators can be used as 'one-off' measures and for repeated monitoring of condition or change. Andersen (1999) urged the need for rigorous evaluations and for 'proponents of indicator taxa to go beyond statements of faith about one's pet organisms, and perform the research that tests their validity'.

Most insect indicator groups are selected initially by a relationship with some environmental characteristic, and it is necessary to determine how robust that perceived association may be. Whilst indicator qualities of terrestrial invertebrates are indeed recognised widely in detecting ecological changes, Andersen et al. (2002) noted that their use was then 'more of a topic of scientific discourse than a part of land management practice'. A far-reaching practical matter raised by Andersen

Table 6.6 The three categories of biological indicators as applied to terrestrial insects (McGeoch 1998), and their major roles

Environmental indicators	Detect a change in an environmental state, or used to monitor changes in an environmental state Categories recognised by some workers: • Sentinels: sensitive organisms introduced as 'early warning devices', or to delimit the effect of an effluent • Detectors: species occurring naturally and may show a measurable response to environmental changes • Exploiters: species whose presence indicates probability of disturbance or pollution • Accumulators: organisms that take up and accumulate chemicals in measurable quantities • Bioassay organisms: selected organisms used to detect presence and/or concentration of pollutants or to rank pollutants in order of toxicity
Ecological indicators	Demonstrate the impact of a stressor on biota, or monitor longer-term stressor-induced changes in biota
Biodiversity indicators	Identify diversity of taxa in a specified area, or monitor changes in biodiversity

et al. is that most land management agencies ignore insects, finding them too intimidating to use because of general unfamiliarity, taxonomic challenges, and vast numbers. This impediment remains difficult to overcome, with development of effective and simple protocols for using invertebrates as indicators a key need. Those protocols should in principle parallel those already widely used for aquatic insects, such as in EPT investigations. In any such sampling programme, the 'trade-off' between simplification and loss of the information available from complete surveys must be considered—and if necessary—assessed.

As a representative context, Australia's diverse ant taxa are used widely as bioindicators in land management and restoration, with patterns of incidence and recolonisation mirroring those in other invertebrate groups. They have received considerable attention at both species and 'functional group' levels. A comprehensive ant survey, usually using small pitfall traps, can yield high numbers of species—in a survey at Mt. Isa (Queensland) discussed by Andersen et al. (2002), 174 species from 24 genera were found. The simplified survey in the same area involved the ant bycatch from the large (bucket-sized; 20 litre) pitfalls linked by 10 m drift fences and used for sampling vertebrates. Sorting of the ants was simplified greatly by considering only the larger (4 mm long and over) ant morphospecies (for a total of 41 species in 12 genera) and noting simply presence or absence of each at each site. This procedure required less than 10% of the effort (cost) of the intensive survey and reproduced most of the key findings of the intensive survey in which all species were evaluated. A suggestion from this outcome was that, rather than continuing to examine the relative merits of different candidate insects as bioindicators, it might be more rewarding to concentrate on making such known useful groups more accessible to land managers and consolidate their use in practical environmental assessments. Once invertebrates become accepted by land managers, attention could then turn to deciding which taxa might be the best to use (Andersen et al. 2002, 2004).

Using dung beetles (Scarabaeidae) in African savanna, McGeoch et al. (2002) distinguished between 'characteristic species' and 'detector species' as indicators. Characteristic species are those with strong habitat specificity and fidelity, whilst detector species span a wider range of ecological conditions. The dung beetles included sets of species that were reliably characteristic of their habitat and, in clearly responding to a change in habitat between states, detector species provided information complementary to that from characteristic species. The commonly used 'Indicator Value' (IndVal) of a species combines its degree of specificity to an ecological state (commonly a 'habitat') and its fidelity (frequency of occurrence) across sampled sites within that state; species with high specificity and high fidelity have high IndVals. Both features facilitate sampling and monitoring as the fundamental need of a bioindicator.

Every insect species signalled as of conservation interest or concern becomes an ambassador for its habitat and the wellbeing of other biota that share threatened resources in that environment.

References

Adler PH, Courtney GW (2019) Ecological and societal services of aquatic Diptera. Insects 10:70. https://doi.org/10.3390/insects10030070

Andersen AN (1999) My bioindicator or yours? Making the selection. J Insect Conserv 3:61–64

Andersen AN, Hoffmann BD, Muller WI, Griffiths AD (2002) Using ants as bioindicators in land management: simplifying assessment of ant community responses. J Appl Ecol 39:8–17

Andersen AN, Fisher A, Hoffmann BD, Read JL, Richards R (2004) Use of terrestrial invertebrates for biodiversity monitoring in Australian rangelands, with particular reference to ants. Austral Ecol 29:87–92

Betrus CJ, Fleishman E, Blair RB (2005) Cross-taxonomic potential and spatial transferability of an umbrella species index. J Environ Manag 74:79–87

Braby MF, Yeates DK, Joseph L (2023) Woodland birds and insect decline. Austral Ornithol 123:255–257. https://doi.org/10.1080/01584197.2023.2233758

Bradshaw CJA, Leroy B, Bellard C, Roiz D, Albert C (et al.) (2016) Massive yet grossly underestimated global costs of invasive insects. Nat Commun 7:12986. https://doi.org/10.1038/ncomms12986

Chowdhury S, Dubey VK, Choudhury S, Das A, Jeengar D et al (2023) Insects as bioindicator: a hidden gem for environmental monitoring. Front Environ Sci 11:11.46052. https://doi.org/10.3389/fenvs.2923.1146052

Collins NM, Morris MG (1985) Threatened swallowtail butterflies of the world. IUCN, Gland

Dettner K (2011) Potential pharmaceuticals from insects and their co-occurring microorganisms. In: Vilcinskas A (ed) Insect biotechnology. Biologically-inspired systems. Springer, Dordrecht, pp 95–119

DIPE (Department of Environment and Primary Industries, Victoria) (2014) IDSC aquatic life subindex, explanatory leaflet, Melbourne

European Commission (2023) Communication for the Commission to the European Parliament, the Council, the European Economic and Social Committee and the Committee of the Regions. Revision of the EU Pollinators Initiative A new deal for pollinators. COM/2023/35 final

Fleishman E, Murphy DD, Brussard PF (2000) A new method for selection of umbrella species for conservation planning. Ecol Appl 10:569–579. https://doi.org/10.2307/2641116

Gao X, Ding H, Xia R, Wang H, Kou Q, Ding S (2020) Developing a modified umbrella index for conservation of macroinvertebrate diversity in Taizi River Basin, China. Water 12(3):857. https://doi.org/10.3390/w12030857

Hall CM (2013) Glow-worm tourism in Australia and New Zealand: commodifying and conserving charismatic micro-fauna. In: Lemelin RH (ed) The management of insects in recreation and tourism. Cambridge University Press, Cambridge, pp 217–232

Hoffmann KH (2017) Insect biotechnology – a major challenge for the 21st century. Zeits f Naturf C 72:335–336. https://doi.org/10.1515/znc-2017-0145

Kaczor M, Bulak P, Proc-Pietrycka K, Kirichenko-Babko M, Bieganowski A (2023) The variety of applications of *Hermetia illucens* in industrial and agricultural areas—review. Biology 212:25. https://doi.org/10.3390/biology12010025

Kietska G, Pryke JS, Gaigher R, Samways MJ (2019) Applying the umbrella index across aquatic insect taxon sets for freshwater assessment. Ecol Indic 107:105655. https://doi.org/10.1016/j.ecolind.2019.10568

Kogan M, Lattin JD (1993) Insect conservation and pest management. Biodivers Conserv 2:242–257. https://doi.org/10.1007/BF00056671

Kritsky G, Smith JJ (2018) Insect biodiversity in culture and art. In: Foottit RG, Adler PH (eds) Insect biodiversity: science and society, vol 2, 2nd edn, pp 869–898

Leather SR (2015) Influential entomology: a short review of the scientific, societal, economic and educational services provided by entomology. Ecol Entomol 40:36–44. https://doi.org/10.1111/een.12207

Lewis SM, Jusoh WEA, Walker AC, Fallon CE, Joyce R, Yiu V (2024) Illuminating firefly diversity: trends, threat and conservation strategies. Insects 15:71. https://doi.org/10.3390/insects15010071

Losey JE, Vaughan M (2006) The economic value of ecological services provided by insects. Bioscience 56:311–323

Losey JE, Vaughan M (2008) Conserving the ecological services provided by insects. Am Entomol 54:113–115

McGeoch MA (1998) The selection, testing and application of terrestrial insects as bioindicators. Biol Rev 73:181–201

McGeoch MA, van Rensburg BJ, Botes A (2002) The verification and application of bioindicators: a case study of dung beetles in a savanna ecosystem. J Appl Ecol 39:661–672

Merritt DJ, Clarke AK (2013) The impact of cave lighting on the bioluminescent display of the Tasmanian glow-worm *Arachnocampa tasmaniensis*. J Insect Conserv 17:147–153. https://doi.org/10.1007/s10841-012-94903-0

Milman O (2022) The insect crisis. The fall of the tiny empires that run the world. Atlantic Books, London

Morimoto J (2020) Addressing global challenges with unconventional insect ecosystem services: why should humanity care about insect larvae? People Nat 2:582–595

Ndondo J (2022) Post-Brexit: the death of edible insect farming in the UK? Institute for a Greater Europe (posted April 27 2002)

New TR (1997) Butterfly conservation, 2nd edn. Oxford University Press, Melbourne

Parsons MJ (1992) The butterfly farming and trading industry in the indo-Australian region and its role in tropical forest conservation. Trop Lepid Res 3(Supplement 1):1–31

Peggie D (2022) Trade and threats on *Ornithoptera croesus* (Lepidoptera: Papilionidae), the Wallace's birdwing butterfly on North Maluku, Indonesia. Treubia 49:85–96. https://doi.org/10.14203/treubia.v49j2.4496

Porto RG, de Almeida RF, Cruz-Neto O, Tabarelli M, Viana BF (et al.) (2020) Pollination ecosystem services: a comprehensive review of economic values, research funding and policy actions. Food Sec 12: 1425–1442. https://doi.org/10.1007/s12571-020-01043-w

Rumbos CI, Athanassiou CG (2021) The Superworm, *Zophobas morio* (Coleoptera: Tenebrionidae): a 'sleeping giant' in nutrient sources. J Ins Sci 21:1–11. https://doi.org/10.1093/jisesa/ieab014
Samways MJ (2020) Insect conservation a global synthesis. CAB International, Wallingford
Scheijen CPJ, Richards SA, Smit J, Jones T, Nowak K (2019) Efficacy of beehive fences as barriers to African elephants: a case study in Tanzania. Oryx 53:92–99. https://doi.org/10.1017/S0030605317001727
van Huis A, van Itterbeeck J, Klunder H, Mertens E, Halloran A (et al.) (2013) Edible insects: future prospects for food and feed security. FAO Forestry Paper 171, Rome
Wilcox BA (1984) In situ conservation of genetic resources: determinants of minimum area requirements. In: McNeely JA, Miller KR (eds) National parks: conservation and development. Smithsonian Institution Press, Washington, DC, pp 635–647

Chapter 7
Need for Understanding and Conservation

7.1 Introduction

Conservation, in any form, depends on awareness of need and belief that remediation is necessary and may be possible. As noted earlier, insects are low in any hierarchy of conservation awareness, and most are simply unseen passengers in efforts to 'save' more popular animals. However, even for the most high-profile (and well-known) non-insect target species, such as primates, whales, parrots, and rare vascular plants, for which public sympathy is readily forthcoming, limited funding, knowledge, and resources may thwart ideal conservation planning, to counter exploitative demand and other interests. Historically, insect conservation has developed from concerns over loss of single species—most notably butterflies—and proceeded along parallel paths and scope to most vertebrate conservation, as a primarily species-focused endeavour. Likewise, formal assessment of 'threat' for invertebrates follows the same criteria devised for those better-documented taxa. At whatever level, actions for insect conservation must be underpinned by robust science, and the three priorities nominated for the European Union (European Commission 2023, Table 7.1) to reverse the decline of pollinators transfer easily to many other insect situations.

It is increasingly clear that effective conservation of insects cannot depend only on single species' focus, but must engage with the formidable problems of conserving 'insect diversity'—largely though addressing habitat quality and continuity and often with natural vegetation and uncontaminated water bodies taken as proxies for finer detail habitat assessments. The underlying faith that protecting the variety of insect habitats from anthropogenic changes will indeed conserve their invertebrate denizens motivates and drives such aspirations. Loss and change of natural vegetation and wetlands to forestry, agricultural conversion, urbanisation, mining, and a range of other pressures (Chap. 3) inflict dramatic changes on 'biodiversity', and the

T. R. New, *Insect Conservation in Australia: Why and How*,
https://doi.org/10.1007/978-3-031-66631-5_7

Table 7.1 The three priority actions recommended by the European Commission (2023) to reverse the decline of pollinators

Priority 1: Improving knowledge of pollinator decline and its causes and consequences
• Establish a comprehensive monitoring system
• Support research and assessment
• Promote capacity building and knowledge sharing
Priority 2: Improving pollinator conservation and tackling the causes of their decline
• Improve conservation of pollinator species and habitats
• Restore pollinator habitats in agricultural landscapes
• Mitigate the effects of pesticide use on pollinators
• Enhance pollinator habitats in urban areas
• Reduce the impacts of invasive alien species on pollinators
• Tackle climate change and other causes of pollinator decline
Priority 3: Mobilising society and promoting strategic planning and cooperation at all levels
• Help citizens and businesses to act
• Promote strategic planning and cooperation at all levels

far less tangible effects of climate change superimpose unknown further stress on these eroded environments—and also on the natural ecosystems that remain.

The twin themes of individual species conservation and diversity conservation are discussed in this chapter, recognising that the 'fine filter' of individual species conservation may be impracticable, and that the 'coarse filter' of community conservation may encompass wider insect diversity and ecological functions, but not necessarily safeguard every species present. In practice, the 'coarse filter' equates to the landscape but with the interposition of a 'mesofilter' level of conservation, in which particular features of that landscape (such as characteristic or peculiar physical attributes or vegetation) may receive particular attention. As Samways (2020) emphasised, this recognition provides three operational scales for insect conservation to be pursued.

At any scale, a defined strategic approach is helpful. The IUCN Species Survival Commission advocates a 'Species Conservation Cycle', as a framework that comprises three main sequential activities under the banners of 'Assess' (compiling data to complete Red List assessments, investigate and discover biodiversity, identify key areas for biodiversity), 'Plan' (developing conservation plans for threatened species and habitats), and 'Act' (implementing conservation plans and policies). Any such scheme can be pursued adequately only through effective partnerships in which all interested sectors are involved and cooperate. Details for needs of any particular insect group can be included in any such scheme to produce lists of priorities and key actions. The scheme for fireflies written by Lewis et al. (2024) exemplifies this structured approach (Table 7.2).

Table 7.2 The priority actions for firefly conservation, drawing on the scheme of the 'Species Conservation Cycle' (IUCN Species Survival Commission) (from Lewis et al. 2024) and exemplifying a sequential approach to conservation that can be a model for other insect groups

Build partnerships
Continue public outreach and education to raise awareness of firefly diversity, life cycles, and conservation needs
Encourage participation in high-quality community science projects that yield verifiable data
Seek partnerships to provide sustainable financial support for firefly conservation efforts
Build capacity and interest in firefly research and conservation by mentoring students and early career scientists
Assess
Expand IUCN Red List assessments across geographic regions to identify which fireflies face the greatest extinction risks and to prioritise the protection of these threatened species
Develop, disseminate, and implement standardised monitoring protocols to track long-term population trends for threatened fireflies
Invest resources in biodiversity exploration in understudied regions, including India, Africa, the Philippines, and South and Central America
Fill knowledge gaps using species distribution models and other approaches to facilitate research on distribution, habitat use, and ecology of threatened firefly taxa
Use IUCN Red List criteria to reassess species at regular intervals to track progress and realign conservation priorities
Plan
Expand existing protected areas to safeguard resources needed to support all life stages of threatened firefly species (e.g. buffer zones along mangrove rivers)
Within established protected areas, ensure that conservation needs of threatened firefly species are explicitly included in management plans, for example:
Reduce light pollution spilling into protected areas
Eliminate broad-spectrum insecticide use
Redesign mosquito control programmes to minimise toxicity to fireflies and other non-target invertebrates
Protect fragile wetland and riparian habitat from trampling (e.g. with boardwalks or fencing sensitive areas off from recreational activities or cattle)
Act
Restore degraded habitat to provide full life cycle support for threatened fireflies
Partner with NGOs and government agencies to secure legal protection for threatened species at state and/or national level
Work with local communities and state, national, and regional organisations to implement policies to reduce light pollution and pesticide overuse that threaten fireflies and other insects
Collaborate with local organisations and land trusts to designate firefly Sanctuaries that provide habitat protection and management for threatened species and areas with high species diversity
Disseminate guidelines and work with local communities, tour operators, and visitors to encourage responsible firefly tourism

Any insect conservation project—whether for an individual species or wider target—must have defined goals, with some reasoned concept of the effort/cost to be devoted to attaining each of these, and a variety of contingencies and clear setting of priorities as elements of a conservation plan. Defining these properly at the start of a project does much to ensure that 'proper' planning has occurred and that unexpected outcomes (including logistic matters such as loss of finance or key personnel, as well as in management) may be compensated. The suggested 'agenda' (Chap. 9) includes many of the issues involved, and Samways (2020) also discussed many of the questions that should be addressed.

7.2 Focal Species as Conservation Targets

As for most other invertebrates, designated insect species are massively underrepresented in conservation legislation—other than butterflies, some larger moths, and a few other charismatic groups such as dragonflies, some stick insects, and members of a few families of larger beetles, most insect groups have almost entirely escaped such formal notice of need, even in the best-known faunas where such need is clearly apparent through losses of natural habitats. Other insect groups are sparsely represented, and public reaction to conserving some poorly understood and more obscure taxa may not be supportive: 'curiosity value' may help overcome this, as may tactics such as allocation of an appealing or resonant common (vernacular) name! This was illustrated well by the threatened loss of two key insect species in their only Scottish location, Auchennines Moss, in 2004. Early publicity against the proposal to destroy the site for landfill mentioned the Bog bush-cricket (*Metrioptera brachyptera*) and *Enteucha acetosae*. The public reacted strongly to support the bush-cricket, but *Enteucha* was largely ignored. However, once further publicity gave it the name of 'Sorrel pygmy moth', strong support was forthcoming—simply, the scientific name had no meaning to local people. The story of this successful campaign is given by Stubbs and Shardlow (2012): *E. acetosae* (Nepticulidae) is one of the world's smallest moths, with wingspan only 3–4 mm—it is the smallest insect to have become the focus of a major and well-supported conservation campaign.

The advisory IUCN Red List is used widely as a global standard for first assessment of species of conservation concern, with the 'threat categories' developed there emulated widely in protective legislation throughout the world. In some cases, the advisory listings of the Red List are transferred unaltered to more formal legislation, conferring priority and obligations for further attention. Many more local 'Red Lists' evaluate national or regional representatives of selected taxa, and the most threatened species designated are often transferred, sometimes uncritically, to formal legislative listing for 'protection' under a national or more parochial Act. Legislative 'listing' implies, and in some cases ideally obliges, attention to conservation and may confer priority for limited logistic support for such activities as the species is in some way 'threatened'. Despite some perceptions that the act of 'listing' a species is 'conservation', it is only the first facilitating step in what may become a long and detailed conservation programme—or, as unfortunately in the case of many listed insects, fail to engender any further support or progress because interest and support are not available. Ideally, it is the first stage of an ordered progress to investigate and, if necessary, undertake adaptive conservation measures.

The sequence of steps for single-species conservation planning after listing has been discussed extensively (New 2009; Braby 2018), and is summarised in Fig. 7.1. Braby (2018) listed six main sequential steps in this process, as shown by numbers in the figure. In many instances, the first step (determining conservation status) will have been the initial grounds for listing or other priority ranking, but additional study may refine this evaluation further—often either revealing that the species is more secure or more threatened than initially supposed. In extreme cases, investigation as part of an action plan may show that the species should not even have been

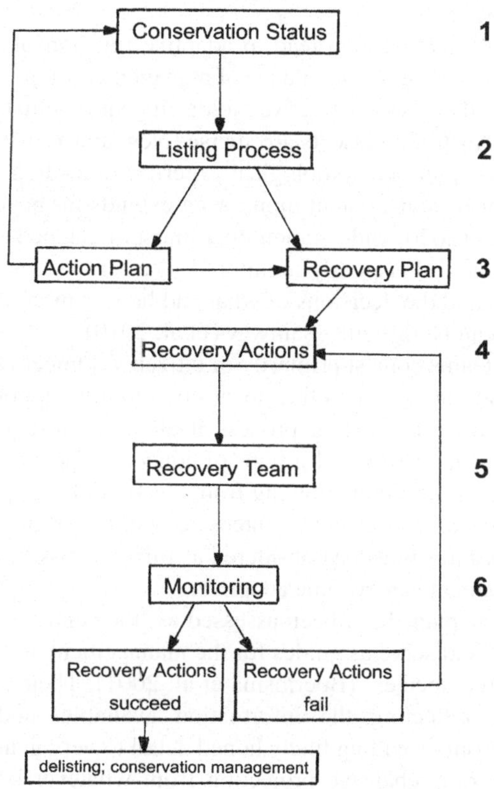

Fig. 7.1 The sequence in steps in conservation management of a threatened insect species (after Braby 2018). The numbers refer to the six main steps noted by Braby, as (1) evaluating the status of the focal species [a step likely to have been done already, but may need re-evaluation]; (2) listing the species for formal legislative recognition, with the threat criteria often established as part of this process; (3) preparing an Action Plan or Recovery Plan, in which the major threatening processes are clarified and measures for mitigation outlined; (4) recommending recovery actions and management options for threat mitigation; (5) implementing those actions through a recovery team; (6) monitoring those actions, with provision for revision and adaptive management if necessary; if management is successful, the species may be considered recovered and a candidate for delisting

listed in the first place, as a rewarding outcome from the listing process. An action plan may simply be a formal legislative obligation following listing, but should be a comprehensive assessment of the species' biology, vulnerability, and conservation needs and recommendations. Those recommendations may include construction of a separate recovery plan, if this is not already embraced as a component of the action plan. This sets out the practical needs and approaches to a conservation programme, with estimated costs and timelines for each step, or theme, and suggestions of 'who does what'; the primary aims include threat mitigation and ensuring habitat security and quality, if necessary through restoration or wider rehabilitation, together with provision for periodic reporting and review. This practical phase should be coordinated effectively by a 'recovery team' of stakeholders and additional expertise as found necessary.

Monitoring of all actions may lead to adaptive revision of actions for greater benefit, or changes of emphasis or balance of management steps, in the hope that the species may eventually be safe. Effective monitoring and evaluation of conservation recovery plans for Australian insects are sparse (New and Yen 1995, 2013), reflecting in part the wider limited entomological expertise in leading conservation agencies. However, poor or non-existent monitoring extends far beyond insects, so that Lindenmayer et al.'s (2020) endorsement that 'most threatened species and ecosystems remain unmonitored or poorly monitored' is a serious concern. Monitoring is a complex process, and the decisions of what and how to monitor and how to interpret the outcomes can be difficult (Samways et al. 2010).

Some recovery teams consist primarily of citizen volunteers and persist through activities of 'Friends groups' or other local community supporters, perhaps with limited guidance. Both the listing process itself and subsequent activities may depend on the enthusiasm of small groups of concerned people. In short, an insect species conservation programme flowing from election of the species as threatened can become an expensive, logistically intensive, and uncertain exercise, for which priorities and procedures must be considered as fully as possible and rendered flexible so that management can become adaptive.

Species' recovery plans have been assessed as 'the central documents available to decision-makers and serve as guides for the management and recovery of threatened and endangered species' (Boermsma et al. 2001). Their content varies enormously in detail, reflecting the information available, and individual plans necessarily range from providing finely honed detail to very general statements, but each should be a comprehensive projection from whatever knowledge has been accumulated. For some taxa a rather broader approach and perspective may be helpful in grouping related species with similar overarching needs into a single plan that can provide general guidance and also indicate the idiosyncrasies of each included species. The hundred or so endemic species of large flightless Orthoptera known as 'Weta' (Anostostomatidae, Raphidophoridae) are New Zealand's paramount flagship insects and of considerable conservation concern, largely due to losses from predation by introduced mammals. A national plan for the group (Sherley 1995) was based on the overlapping biologies of the included species, but also included appraisal of each individual species to illustrate the variety of approaches needed for even closely related insects.

Likewise, Douglas (2003) included five species of sun-moths (Castniidae) found in Victoria in an embracing report, in which their common and individual needs were discussed. Defining the 'common needs' for an insect group, however tentatively and incompletely, may give valuable initial 'clues' for planning conservation of its individual members, or for the group's representatives in a given conservation exercise. The sun-moths, for example, all depend on native grassland so they share need for this vulnerable environment.

Examples of recovery plans or action plans for insect species reveal the array of possibilities and desirable measures that can arise, in many cases with priorities indicated, so that actions can be 'ranked' for benefit or urgency. Later review may reveal little or no progress. The Action Plan for the narrowly endemic Mt. Donna Buang wingless stonefly (*Riekoperla darlingtoni*, p. 148) in Victoria (Ahern et al. 2003) listed the major conservation objectives as a base for a series of intended management actions in four categories (Table 7.3), with a total of 14 actions, each with one or more agencies listed as responsible for pursuing it—but with little practical action flowing from these. The table simply indicates the considerable variety of themes that may arise in any such species treatment and that vary enormously in their practical feasibility. Essentially, such appraisals are a well-considered 'wish list' of the conservation measures designed to benefit a focal species. However, in 2023 pressures from proposed recreational development (for a cycling track near the stonefly habitat) exposed the vulnerability of the stonefly to habitat change from disturbance; it was by no means secured from any previous actions, but documentation of its very local distribution and restricted habitat was later instrumental in preventing the track development from proceeding.

Any detailed management plan for an insect species can thus provide useful guidance for (1) related species or (2) other insects in the same or a similar habitat or under similar threats. However, every insect species is different—and the details applicable to one species may not transfer uncritically to even its closest relatives. It is almost always necessary to undertake original investigations on a previously unstudied species and, sometimes, for a previously studied species in a different environment.

A major limitation is that most insects have not been assessed for conservation status allocation, so in the Red List are 'Not Evaluated'; however, many of the evaluated species are necessarily allocated as 'Data Deficient' because little is known about them. Nevertheless, 2345 insect species were listed on the IUCN Red List by 2012 as 'Vulnerable', 'Endangered', or 'Critically Endangered' (i.e. the three categories classified as 'Threatened': Fig. 7.2), already a total daunting to accommodate in practice—and amongst which most will not be sufficiently supported from current finances and expertise to meet their conservation needs. Numbers vary somewhat as assessments become more numerous and more refined, so that by 2021 2270 of the then evaluated 12,100 insect species were assessed as Threatened. Progressive status evaluation of insect groups using IUCN Red Data

Table 7.3 The management actions listed for the Mount Donna Buang wingless stonefly, *Riekoperla darlingtoni*, abbreviated from listing by Ahern et al. (2003)

Research and monitoring
1. Identify areas of potential habitat of the stonefly within at least 5 Km of Mt. Donna Buang, based on altitude and vegetation
2. Undertake surveys to establish distribution and relative abundance of the stonefly at sites of occurrence
3. Initiate and support research into an efficient means of estimating the size of breeding populations and of monitoring population trends at individual sites of occurrence and monitor selected sites annually for several years
4. Encourage and support research into the ecology of the stonefly
Habitat protection
5. Encourage development of park management strategies which avoid further expansion of visitor facilities, especially car parks or other landscape modifications
6. Ensure that Park management procedures incorporate appropriate measures to protect the stonefly and its habitat on an ongoing basis
7. Monitor and, if necessary, mitigate silt input and erosion processes along water courses known or likely to constitute stonefly habitat
8. Ensure, on an ongoing basis, that all Park management staff are aware of the localities of occurrence of the stonefly and of potential threats to the stonefly or its habitat
9. Through signage or other interpretation encourage the public to promptly advise Parks Victoria rangers of any chemical spills likely to enter local run-off and affect downstream Park values
Public values
10. Devise and implement an interpretation programme at Mount Donna Buang and nearby information centres to inform the public of the features and significance of the stonefly, emphasising opportunities for public assistance through environmental care and support for management processes
11. Investigate prospects for involvement of volunteers with relevant expertise in stonefly population monitoring or other coordinated research or management activities, providing that their involvement will in no way threaten the taxon
Other desirable management actions
12. Ensure lodgement with the Museum of Victoria of all outstanding specimen material from past studies of the stonefly
13. Determine appropriate vegetation buffer width to protect the species, based on the area over which individual stoneflies range
14. Encourage public and government cooperation to help reduce gas emissions likely to contribute to enhanced greenhouse effect and global warming

Book criteria demonstrates previously unexpected levels of threat: detailed evaluation of the European Orthoptera (Hochkirch et al. 2016), for example showed that more than a quarter of the 1082 species were threatened. These included 231 of the 739 European endemic taxa. By 2021, 314 of the 890 European species of hoverflies (Syrphidae, a significant group of pollinators) were assessed formally as threatened, as were 16% of the 6016 global species of Odonata (dragonflies and damselflies).

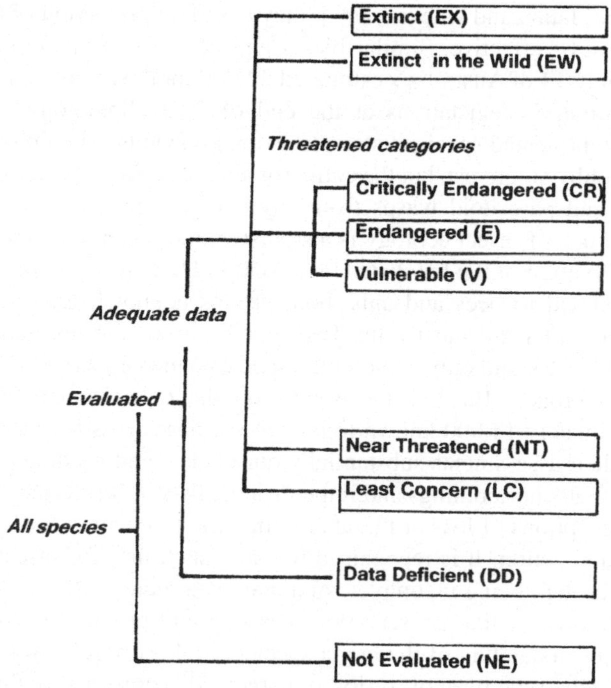

Fig. 7.2 The scheme for species status assessment using the IUCN Red List criteria, showing the categories to which a species may be allocated. 'Risk of extinction' for evaluated taxa increases towards top of the page (based on figure from IUCN 2012; guidelines for using categories are explained in IUCN 2017)

One approach to overcoming the logistic impossibility posed by the vast numbers of insect species needing assessment has been to examine a large randomly selected sample deemed to represent the larger taxon. Odonata were the first group to receive this treatment globally (Clausnitzer et al. 2009). A sample of 1500 randomly selected species of dragonflies and damselflies—more than a quarter of the 5680 species then known—across all biogeographical regions was assessed using the IUCN Red List criteria, with the expressed hope that the same species would be assessed at future intervals to indicate changes. Only about 10% of the species were found to be threatened with extinction. Many of these were in the Indo-Malayan and Australian regions, and island endemics were also amongst the most threatened species.

Likewise, only a tiny proportion of insects have been assessed for inclusion in protective legislation. Thus, in Europe, only 123 of the estimated 105,000 known species are currently protected (Leandro et al. 2017), with strong bias towards those that are large, better known, and more widespread. Leandro et al. asked the very reasonable question of whether those listed species were indeed representative of the entire insect fauna and whether any policy based on protection of this 0.12% of the fauna could be effective. Similar biases and poor representation apply widely elsewhere—only 14 of Australia's estimated 22,000 moth species were listed under any of the country's legislations at the end of 2020. Taxonomic bias towards Lepidoptera in protected species lists reflects the knowledge accumulated over 1–2 centuries of hobbyist interest, but the entire (or almost entire) absence of vast groups such as flies and parasitoid wasps from any national, regional, or global listing ensures that much of insect ecology is also disregarded, and may not be addressed through conservation of the few listed taxa. Within Hymenoptera, protective listing is largely confined to bees and ants, both groups acknowledged as functionally important. The latter are mostly the Red wood ants of the northern hemisphere (Balzani et al. 2022), and only some of the species had been assessed for the IUCN Red List. More broadly, Balzani et al. pointed out that (as of January 2022) only 149 of the approximately 14,000 known living ant species across the world are on that list, all with their assessments substantially out-of-date and needing revision.

Whatever their shortcomings and imprecisions, lists of 'protected species' based largely on perceptions of loss or threat are often the foundation for local conservation interest and action. It is recognised that any such lists for insects have major deficiencies through being incomplete and that even many of the species listed are very poorly known so that conservation planning and progress must be based on very limited understanding of the target species' biology and distribution. Impetus for listing reflects some form of 'rarity' or perceived decline and/or threat, inducing risk of extinction. Most candidates are ecological specialists with very restricted distributions, such as local endemics, and many are nominated from concerns raised by individuals or local community groups. Focus on single species, even with considerable gaps in knowledge and uncertainty over their true needs, provides a 'tangible' focus for conservation—and largely emulates the major approach used for vertebrates. In general, it is far easier to advocate for a named and distinctive 'Butterfly A' or 'Dragonfly B' than for more nebulous entities such as ecological communities; for most recipients, simply promoting an insect species is itself unfamiliar—perhaps even risible. And, perhaps more seriously, the great majority of conservation managers are not entomologists, and enter an intimidating, bewildering, and daunting new and complex arena in adopting or becoming committed or directed to insects. Experiences with vertebrate conservation do not always transfer easily to tiny unfamiliar and rare insects—even that the larvae and adults of insects with a complete metamorphosis are essentially different biological entities, usually with different habitats, behaviour, feeding habits, and other resource needs, with a strong pattern of seasonal development and apparency may represent 'a new world', for which navigation advice on the voyage of discovery is very limited.

The biology of 'Butterfly A' or 'Dragonfly B' and their distribution may be very poorly known—but also be far better known than for most of their relatives. Each focal species might be recorded only from few adult specimens or records from one or few sites. Systematic surveys for incidence and abundance are unlikely to have been made, early stages may not be recognisable or associated with adults, flight period can only be inferred from existing records, butterfly food plants may be unknown, and other gaps in knowledge compound to reveal a level of complexity and knowledge insufficient to pursue meaningful conservation other than in very general terms—and usually in marked contrast to more familiar and larger vertebrates. For many such insects, practical observations are limited by their scarcity, and by lack of knowledge of where and when they occur, and how to find them or even recognise them. Small rare insects are difficult to study.

Limited available information may indeed be misleading. For example, the Golden sun-moth (*Synemon plana*, Castniidae) is now known to be much more widely distributed in native grasslands on Australia's south-east mainland than earlier supposed. Its previously assumed scarcity was in large part an artefact; assessments were based mainly on visual sightings of the diurnally active male moths (females fly little), but males are now known to fly for only a few hours in late morning-early afternoon, with weather strongly influencing their activity. Surveys at other times of day, or if temperature or cloud cover were unsuitable, failed to detect the moth (Gibson and New 2007). Investigations under the right conditions and timing revealed the moth at many other sites. A further complication is that individual moths live for only a few days, but the flight season at any site extends over around 2–3 months, so that emergences continue over that period. In addition, it is still not clear whether the life cycle (larvae feed underground on grass roots) takes 1, 2, or even 3 years, so that the moth emergences in any 1 year may simply be a cohort of a larger unassessed population.

Accurately studying any listed insect may require wide consultation to uncover all available background information and form a valid and constructive approach. Incorporating entomological expertise and seeking expert advice is recommended strongly, not least to avoid misplaced effort. And, in addition, any such study is likely to reveal biological idiosyncrasies that render the species unique and, for many, demanding some deviation from any formulaic conservation plan. The details above for the Golden sun-moth indicate the subtleties that may influence field interpretations and observations, and the problems that can arise from not doing further research but which may become the foundation of far-reaching policy.

A further consideration is also difficult to address. Most single-species conservation is in some way 'site-' or 'habitat'-focused, with conservation management—whether for species persistence or translocation/introduction (p. 124)—seeking to emulate, sustain, or restore the conditions on some 'reference site' or knowledge of the species' needs. Some rare species, however, rather than thriving, may simply be 'hanging on' at sites on which conditions are suboptimal, even marginal, for persistence. Management may inadvertently seek to emulate those marginal conditions rather than those that would substantially benefit the focal species. Another ecological situation can also provide ambiguity, reflecting incomplete knowledge of the species' population structure. It is sometimes unclear whether a local loss from a site (extirpation) is really the end of a 'population' or whether the focal species has a wider form of population structure, as a 'metapopulation'. This often manifests as a continuing series of local extinctions and re-colonisations of isolated habitat patches in a wider landscape—so that any particular site may or may not host the species at any given time, but its absence may not be significant.

A major priority, often noted as a general recommendation or obligation in a listing approval for most listed insect species, and reflecting wider uncertainty over how to proceed, is the deceptively simple task of finding out more about the species population dynamics, biology, and susceptibility as a basis for evaluating real conservation need and designing a suitable management programme. Many such approvals lead to (or in some cases, formally oblige) production of a management plan or similar document, with timing, demands, and detail varying considerably across legislations and species. They may also allow for revision of listing status pending review at intervals. Increasing the knowledge of a supposedly rare and threatened insect species is often a protracted and intricate process, involving seasonal field work and interval sampling for monitoring over extended periods, by methods that may be initially unfamiliar to those undertaking the work. In practice, many such demands have not been met by imposed deadlines, and some listed species are 'in limbo' pending completion of such plans. Two immediate demands arise—(1) in some way assessing the distribution and vulnerability of the species, and defining its biology and life cycle in sufficient detail to understand those constraints, and, in many cases (2) monitoring the species to evaluate changes in distribution and abundance and clarify population structure.

In contrast to many insect taxa, because butterflies and dragonflies are amongst the best-studied insect groups some background biological information on species related to or co-occurring with the nominated taxon is likely to be available, from either published articles or by consulting local naturalists. 'Butterfly A' or 'Dragonfly B' might be the fortunate beneficiaries of such transferable knowledge, so that a tentative 'working hypothesis' for conserving the target species can be assembled for initial evaluation. By contrast, less popular entities such as 'Barklouse C' would almost certainly lack any such support, be even more inadequately known, and face a formidable burden of proof to justify their status and any conservation investment. Barklice (Psocodea) are amongst the 'meek inheritors' of poorly studied insect groups whose ecological significance is unknown in any detail and for which the situation is unlikely to change markedly in any way that would enhance pragmatic grounds for conservation.

For many insect groups—and even for some members of the best-known taxa—accurate species identifications may be difficult, and can be confirmed only by a specialist versed in the small features used to distinguish species and aware of the natural variation that occurs. Those specialists are themselves often rare, with the consequence that many insect groups are massively understudied; some are essentially 'orphaned' through want of attention, and progress on understanding their systematics and biology remains slow. Any purported conservation need may thus be difficult to justify, and many insect groups (predominantly those in which individuals are small) are simply 'put in the too-hard basket'.

Extending the lists of insect species deserving conservation assessment is in principle desirable in drawing attention to their needs and for basic study. However, even for the 'well-known' British insect fauna, massive practical difficulties occur, and these are magnified elsewhere in the world. Morris and Welch (2023a, 2023b) noted three major reasons why increased listing in Britain could prove counterproductive, as: (1) a very high proportion of species are less than 5 mm long and require microscopic examination for reliable identification—the pool of people capable of undertaking that work, let alone monitoring, is tiny and inadequate for the task; (2) the difficulties of any 'licensing', or permit system, in providing sound credentials, could hamper the development of interests and of a new generation of specialists; and (3) if permissions to sample insects become more onerous, less activity from specialists, together with fewer surveillance data, will render it more difficult to place available records in a context that will distinguish declines from lack of data. The outcomes of increased 'species listing', although often advocated as vital to gain awareness of insect conservation needs, are a serious consideration.

For the many species that are not well known, the proof to gain listed status can be substantial. Australia's EPBC Act requires assessment against five criteria, most demanding quantitative assessments. Problems are illustrated by considering the reasons for rejecting the nomination of the Small Orange-spotted sun-moth (*Synemon discalis*) some years ago (DEH 2007) as a species ineligible for listing. Essentially, the available information was too incomplete for it to be accepted, as follows: (Criterion 1. It has undergone, is suspected to have undergone, or is likely to undergo in the immediate future a very severe, severe, or substantial reduction in numbers) 'Insufficient data available to assess the species against this criterion'; (Criterion 2. Its geographic distribution is precarious for the survival of the species and is very restricted, restricted, or limited) as for 1; (Criterion 3. The estimated number of mature individuals is limited to a particular degree and (a) evidence suggests that the number will continue to decline at a particular rate; or (b) the number is likely to continue to decline and its geographic distribution is precarious for its survival) 'there are no quantitative data available to assess the species against this criterion'; (Criterion 4. The estimated total number of mature individuals is extremely low, very low, or low) as for 3; and (Criterion 5. Probability of extinction in the wild) also as for 3.

Listing any insect species for formal recognition is a responsible action, not to be taken lightly, and demanding knowledge of the species' status and risks of decline or loss. It is, however, important to differentiate between 'listing' and 'conservation'. Allocating a category of threat using the IUCN criteria is a statement of 'risk of extinction' based on the best available evidence. Selection of such a species for conservation incorporates many other values, such as feasibility, cultural or phylogenetic importance, wider importance of its habitat, competing or complementary interests, local priority, and so on. A species may vary considerably in abundance and risk across a broad range—such as in different countries of Europe, as shown by Collins (1987)—and transcend a series of different legislations with different conditions. In Australia, the eight States and Territories each have their own legislation, a situation that has historically sometimes led to inconsistencies over reaching consensus on needs. Greater unity is now present through the Commonwealth's Environmental Protection and Biodiversity Conservation Act.

A 'threatened species list' is often the only formal basis for selecting taxa for further attention. Not only is a listed insect likely to gain priority for scarce conservation support, but also to deprive other equally deserving unlisted taxa of such support. Under some legislations, any non-listed insects are essentially excluded from official support. Each listed species thereby becomes a 'flagship' for wider conservation interests and, perhaps, an 'umbrella' for numerous other taxa in the same area or habitat. Each listed species will also have its specific idiosyncrasies and the almost universal comment in listing documents, which in some form recommends further study and clarification of status and needs, may mask formidable complexity. Many insects are deceptively difficult to study, and doing so to meet the important needs to elucidate species' features such as life history and seasonal development, population size and structure, distribution, and vulnerability commonly involves long-term effort drawing on the best available knowledge and methodology. Much of the methodology and the biological features displayed are likely to be unfamiliar to non-entomologist conservation managers. Insect species conservation must generally proceed with very inadequate information to guide it, and initial measures are necessarily general, to address the most obvious and urgent issues arising. A series of immediate 'no-regret' measures for insect conservation were proposed by Harvey et al. (2020) as 'beneficial to society and biodiversity even if the direct effects on insects are not known as of yet': hence the epithet 'no-regret' (Table 7.4). These can then lead to mid-term and long-term actions as information accumulates and enables more detailed focus on the taxon or environment of concern.

Broadening from single-species considerations to wider perspective is increasingly recognised as a (perhaps the most!) constructive way to advance. It rests on reasoning that high insect diversity ensures that any formal listings of individual species will be of only a small subset of deserving species, but these will still by far outstrip the resources available to pursue their conservation. In short: (1) most insect species will not be assessed properly for individual conservation need; (2) even the few listed species will not have their needs addressed properly from currently available support; (3) those species represent a taxonomically biased subset, itself highly incomplete, of insect life with most orders not represented; (4) many key ecosystems lack any listed 'insect ambassadors' so their conservation is not enhanced; (5) expanding species listings on the currently predominant 'ad hoc' basis is not likely to overcome these impediments, so (6) other approaches are needed to effectively conserve insect diversity.

Table 7.4 The 'no-regret' measures proposed for immediate action in insect conservation (Harvey et al. 2020) (not in any priority sequence)

Increase landscape heterogeneity in agriculture
Reduce light, water, and noise pollution
Phase out pesticide use and replace with ecological measures
Reduce imports of ecologically harmful products
Avoid and mitigate alien species introductions
Conservation of threatened species
Enhance restoration and conservation programmes
Education for awareness, citizen science, and capacity building

Many of the species listed are highly specialised, occur in low abundance, and are very localised, all features that render them difficult to study without causing further harm by take or disturbance. Nevertheless, as Taylor et al. (2018) recommended for Australia, selectively increasing lists of threatened insects to include representative species characteristic of more of the nation's major biomes ('bioregions', many of which have no listed species as ambassadors and to demonstrate their vulnerability) could help this wider conservation need (p. 168). A considered rationale for selection of listing species could lead to more efficient deployment of support. Individual species focus—the so-called 'fine filter' approach to conservation—is likely to persist but cannot remain the mainstay for conserving insect diversity. It is very expensive, and past experience has demonstrated amply that the approach depends to a large extent on the dedication of volunteer aids to augment the limited official support. Logistics dictate that the broader perspective of focusing on insect diversity as a 'coarse filter approach' to insect conservation must receive increasing attention and become a major policy and practical focus. The single-species approach is queried increasingly as its limitations are more fully acknowledged—as Morris and Welch (2023a, 2023b) commented 'extending lists of protected [insect] species is a vertebrate ecologist's response to nature conservation and illustrates the ongoing, yet long recognised, problems of taxonomic bias'.

Single-species conservation for insects is most commonly pursued on sites where the species occurs, however parlously, as in situ conservation. In some instances, however, this may be replaced or augmented by ex situ measures involving translocations or other introductions of individuals to historical or more novel sites—such as those outside the current range. Both are based on a series of basic principles and are outlined below. Whichever focus is adopted, an 'advisory group' should usefully include representatives of all stakeholder groups whose viewpoints and advice may constrain or guide conservation planning and any later management.

7.2.1 *In Situ* Conservation

Once an insect is acknowledged or suspected to need conservation, by listing or other priority, its status usually needs fuller exploration, with the purpose of conserving it where it naturally occurs. It may be necessary to explore whether the species is truly resident and breeding on a focal site, or whether it is a transient visitor. In many cases it may be known from few records and only from one or few sites, and assuring security of that occupied habitat is a primary need—with past, continuing, or imminent habitat destruction often the most apparent threat, but all management pointless if threats to the site(s) continue. The last can require immediate actions to attempt to delay or halt destruction pending further investigation of the insect—steps likely to be opposed strenuously as pressures mount for increased urban or industrial developments, road modifications, mining exploration, and other major changes. It may be feasible to gain a temporary halt to development, or moratorium, in order to enable further investigation. Site security may also involve need for change of land tenure—for example when the species is on private land and may be vulnerable to a change to less sympathetic ownership.

7.2.2 *Ex Situ* Conservation

The deliberate movement of individuals of priority species to different sites, broadly translocation, is a widespread tool in conservation and has been used for insects mostly for Lepidoptera, Coleoptera, Orthoptera, and Odonata, in several distinct contexts, with fewer attempts for several other orders (Bellis et al. 2019; Nason et al. 2021). The overall aim is to establish or assure a self-sustaining population that can persist and reproduce in a secure environment, through a broad series of guidelines proposed by the World Conservation Union (IUCN 2013). The individuals involved may be from captive reared stock or transferred from existing field populations (with due regard not to weaken the donor population by reducing its numbers to a critical level and increase its vulnerability), with purposes including (1) reinforcing small populations by increased numbers and genetic variety, so enhancing population viability; (2) re-establishing a population in part of the historical range, perhaps after site rehabilitation and protection from further disturbance; (3) establishing new populations beyond the historical range by assisted colonisation, sometimes most important for species with low dispersal powers (such as wingless grasshoppers); and (4) 'salvage', the rescue of individuals from a doomed site for release elsewhere. The last may become urgent and extreme, sometimes as a response to unforeseen disturbance, but greater notice may also be available for longer-term planning. Two cases involving Australian Lepidoptera illustrate these contrasting scenarios:

(1) A population of the myrmecophilous Purple copper (also known as the Bathurst copper or Lithgow copper) butterfly (*Paralucia spinifera*, Lycaenidae) in the path of a road realignment project in New South Wales was discovered only after the project was well advanced. The project was halted to allow consideration of rescue. Potted larval food plants were sunk in the ground at the donor site and later transplanted with the host ant (*Anonychomyrma itinerans*) to a nearby new site that had already received nursery-grown food plants. Caterpillars on the moved plants were augmented by those collected by night from the plants in the area to be destroyed, and these were released on the new site. Success of this rescue operation (Mjadwesch and Nally 2008) was possible only through cooperation of all parties involved, and because conservation of the butterfly was a well-supported local issue.

(2) A population of the Golden sun-moth (*Synemon plana*, Castniidae) in central Victoria occurred in grassland through which a major water pipeline was to be installed to bring water to Melbourne from the Goulburn River catchment. The pipeline route was planned well in advance of work beginning. Rescue was attempted by removing sods of grass likely to contain the subterranean larvae of

the moth and inserting them elsewhere, to allow the larvae to develop without direct disturbance. In the Australian Capital Territory, attempts to translocate caterpillars directly by recovering them from soil rips and inserting them into holes near foodplants at a new location have met with mixed success.

It may be necessary to breed a species in captivity pending reintroduction to the wild, either to build up numbers in a protected environment or because a suitable release site is not yet available. Prolonged rearing is deemed generally undesirable, because of possible genetic degeneration such as by inbreeding, but a very small founder population may be unavoidable for a rare insect. One example is of the Lord Howe Island stick insect (*Dryococelus australis*, listed as Critically Endangered by IUCN in 2017), for long believed extinct (most likely from predation by introduced rats) before discovery of a tiny population on the remote Ball's Pyramid. A pair was transferred to Melbourne Zoo, where captive breeding commenced (Honan 2008). As of early 2023, about 25 thousand stick insects have been reared (Pearce-Kelly et al. 2024), with populations in due course shared across several other zoos in various parts of the world as an additional buffer against calamity. The eventual aim is to reintroduce *Dryococelus* to Lord Howe Island and neighbouring islets once a rat eradication campaign has been completed—hopefully within a few years—and that major supposed source of mortality no longer exists.

Many translocations and reintroductions are to local sites with relatively recent records of the species' occurrence, but larger scale cases also occur. The classic cases of successful introduction of European subspecies of the Large copper (*Lycaena dispar*) and Large blue (*Maculinea arion*) to Britain after loss of these butterflies there and the intricate research needed to maximise chances of success illustrate the detail necessary to understand the species' biology and needs.

Because any translocation or similar exercise is to some extent experimental, it should be monitored carefully, preferably over an extended period. Although there is no standardised definition of 'success' of a translocation, appearance of the next generation after introduction is often accepted for this with chances of success increased by a larger number of translocated individuals (Bellis et al. 2019). Small numbers of insects are often elusive and difficult to detect, and the outcome may not be clear for at least several generations. The wingless morabine grasshopper *Vandiemenella viatica* has lost much of its former habitat around Melbourne but 1 year after experimental translocations to small open urban areas was recovered at 28 of the 36 receptor sites (Yagui et al. 2024), demonstrating that such generalist species may be amenable to movement procedures and that small urban patches may be suitable for this.

A population of the dragonfly *Leucorrhinia dubia* (Libellulidae) introduced (by transferring larvae) to artificially created bog pools in Czechia was monitored for 15 years by Dolny et al. (2018), over which it became dominant and the resident odonatan assemblage did not change substantially. As in many aspects of insect conservation management, long-term monitoring by standardised methods is an important aid to understanding—but is rarely practicable over sufficient periods; changes in personnel, funding, and local priorities can preclude continuity.

Another precaution is to ensure that details of all translocations and other 'movements' of species are recorded, in order to avoid future misinterpretations of distribution; in the past simply releasing surplus specimens reared by collectors in places far from their recorded habitats has led to confusion in plotting distributions. In extreme cases, in the heyday of collecting, unscrupulous importation and release in southern England of Lepidoptera rare in Britain but more common on continental Europe allowed collectors to claim capture of a 'British specimen' with considerable commercial value (Allan 1943), and for the 'discoverer' to be rewarded generously for alerting the collector involved! Release of surplus insects should be undertaken as close as possible to their origin, to preserve genetic integrity and avoid undocumented mixing of genotypes from different populations.

Considerable information relevant to conservation translocations and releases of insects can be derived from literature on classical biological control, whereby agents such as parasitoid wasps and predatory insects, often mass-reared, are released to suppress pests, and monitored to determine their impacts.

7.3 Approaches to Conserving Insect Diversity

As noted above, individual species conservation focus is attractive and popular, but resource-intensive, often long-term, and logistically difficult to sustain. It is impracticable to construct individual recovery plans for every deserving insect species. And even short lists of threatened insect species (p. 158), so often given priority or commitment for further action or investigation, by far outstrip the resources (from finance to expertise) available. Those lists are unlikely to be fully representative of need, are taxonomically biased, and overlook numerous species that may be equally deserving. Many of those species are poorly known, apparently highly localised, perhaps ecologically specialised, and lack any advocacy for conservation—because their needs cannot be appraised from the scant information available. Simply, and however appealing, this approach to insect conservation in regions with diverse and poorly documented faunas is inefficient, and usually impracticable. Even in the wealthiest parts of the world in which conservation need is acknowledged readily, the funding and expertise needed to conserve the limited number of listed insect species are likely to remain inadequate and their deployment for greatest reward controversial.

Wider perspective and broader approaches to insect conservation are clearly necessary. These devolve on combinations of habitat protection and enhancement and threat prevention or suppression in both natural and anthropogenic environments. Surveys of conservation needs for higher insect taxa can lead to general recommendations to increase their protection, and be based firmly on the fundamental needs of the group. For Odonata, Clausnitzer et al. (2009) noted that 'Specific conservation action plans for every threatened dragonfly can neither be achieved nor realised' and suggested a number of general conservation measures that could embrace needs of many of the most needy species, those of lotic tropical forest habitats (Table 7.5).

Table 7.5 The general conservation measures for threatened Odonata, especially species of lotic habitats in tropical forests (abbreviated from list by Clausnitzer et al. 2009)

1. Prevention of any further deforestation, with forests regarded as a natural resource with high conservation value
2. Afforestation with indigenous trees
3. Fire control
4. Leaving a strip of natural riparian vegetation of at least 20 m on either side of streams and rivers in cultivated and mined landscapes
5. Water pollution control measures, such as from sewage works, mining run-off, and factory effluent treatments. Restricting and controlling amount of insecticides and pesticides
6. Protection of watersheds, floodplains, water retention sites, swamps, etc., against adverse impacts of damming and other alterations

Such general measures clearly have far wider relevance, and parallels can be designed for many other insect taxa.

Threat management is the central theme of conservation worldwide, but recognition that insects need specific tailored management programmes in order to achieve this is sometimes hard to communicate. The ideal of insects being adequately conserved in secure suites of protected areas that are sufficiently large and well-buffered against disturbance, represent the full range of local biotopes, are well-connected across a regional landscape, and are managed to ensure that their values for insect diversity are sustained is utopian—but remains an objective in policy design in conjunction with ameliorating the drivers of declines (Duffus and Morimoto 2022).

However, a wider perspective is needed for the landscape of 'unprotected areas' and including much of the agricultural and urban estates critical for insect (and other) conservation. Many lessons of practical management for insect diversity there have wide applications and potential for modification for individual circumstances.

7.3.1 Insects in Protected Areas

Although vitally important as sanctuaries for conserving high-profile vertebrates, very few national parks or other designated reserve areas have been erected specifically for insects, which thereby become largely passive 'passengers' with priorities lower than for any designated target taxa. Their neglect in this context means that insects (and other invertebrates) have no profile amongst policy-makers and conservation managers/practitioners (Morris and Welch 2023a, 2023b) whose activities may govern their future survival. One notable exception is for dragonflies in Japan, where their cultural interest has led to a number of dedicated reserves to conserve them. And in Britain, as in other parts of Europe, popular interests in Odonata and Lepidoptera have fostered establishment of some reserves. Parallels remain scarce elsewhere.

Most protected areas, however, have other designated priorities. Many larger parks have multiple uses, including ecotourism and other recreation, and management may give such activities priority—not least to raise funding for conservation work within them. Those activities may unwittingly increase threats to lesser-known biota. Slashing or mowing of grassland to increase visibility of larger mammals for tourists, for example, may directly erode low vegetation or other insect resources. However, the widely suggested maxim that 'insects will be safely conserved under the umbrella of protected large mammals (or other vertebrates) in reserves so that more specific conservation is unnecessary' is largely untenable, notwithstanding the opportunistic—and in some cases manipulable—co-existence of insect diversity in those areas which may then function as refuges. The factors determining insect incidence in particular habitat patches are often unclear. Many insects are restricted to particular vegetation associations and/or seral stages, so that their occurrence in any place may be transient. Those places may be very small, mirroring the natural distributions of some narrow-range endemic insects, and intrinsically vulnerable through edge effects and other disturbances. However, in many cases 'leaving insects alone' even in a securely protected area may not suffice to sustain them as conditions change. Although not including insects in their analysis of the safety of the 7784 threatened species in the IUCN Red List (as of March 2021), Bolam et al. (2022) found that more than half (57%) of those species needed active targeted management to reduce their chances of extinction. Australia figured highly, as one of the countries in which high numbers of species each need such focused attention to aid their recovery, in conjunction with threat suppression, and irrespective of whether or not they occur in protected areas. The promotion of 'havens', as areas (islands and fenced mainland areas free from introduced predators) for management and protection of Australian vertebrates (Woinarski et al. 2023), may also benefit co-occurring insects and other invertebrates, as a series of areas already acknowledged to have conservation significance.

Many insects have been reported from National Parks and other reserves but for most of those species little or nothing is known of their status, roles, and vulnerability there—they are simply 'records of incidence', many of them from short-term surveys rather than any systematic searches. Even 'species lists' from major protected areas are likely to be highly incomplete, and 'insect inventories' for most such areas remain unattainable. Discovering systematically what insects occur in protected areas, and so the values of those areas as harbours for insect life in ever-changing landscapes, would be a major stimulus to documenting insect biodiversity and what of this might be amenable to conservation without further designated land/waterbody protection. For Australia, Sands and New (2003) noted that even butterflies were very incompletely reported from National Parks, but that this situation could be remedied by facilitating interest from hobbyists, rather than deterring their interests by complex needs for permits. Earlier, Sands and New (2002) suggested that rare butterflies recorded from national parks should be ranked as less threatened than species known only outside protected areas, so presuming (without evidence) that some 'protection' would indeed occur. For most other insect groups, no such presumption can be made. However, incidence of a 'listed species' in a protected area may be taken to reduce the need for a formal species conservation plan. The species profile accompanying the listing of the Alpine stonefly (*Thaumatoperla alpina*) as 'Endangered' (in 2011) noted 'Recovery plan not required, half of the known populations occur on protected land within the Alpine National Park and the proposed conservation advice provides sufficient direction to mitigate threats at this stage'. That advice, following from a series of threat abatement actions, listed actions to mitigate threats including habitat loss, disturbance and modification, animal predation, and fire, as well as raising awareness of the stonefly. All are possible within a protected area, but depend on resources and goodwill to progress: as 'advice' rather than formal recovery actions, obligations, and priority may appear lower.

Accounts of the values of protected areas for insects vary greatly. Some of the most significant declines of insects reported have been based on surveys within protected areas (Hallmann et al. 2017; Wagner 2020). A number of surveys comparing incidence of selected insect groups within and outside protected areas also display concerning trends. In a study that emphasised the need for long-term data, Rada et al. (2019) showed that butterflies were richer in the European Natura 2000 network of protected areas ('N2000', one of the world's most important and largest conservation networks, covering at present around 18% of the European Union land area) in Germany than elsewhere in the landscape—but that declines in richness in and beyond N2000 protected areas were similar. Both showed losses of around 10% of species over 11 years from 2005. Despite the clear value of those protected areas, they did not prevent or even slow butterfly declines, a circumstance that Rada et al. suggested might reflect insufficient targeted management, or management for organisms other than butterflies. Also in Germany, the dearth of insect-focused management regulations for protected areas—so that insects are largely neglected in conservation legislation and planning—was amongst the concerns raised for effective insect conservation in nature protected areas (Kothe et al. 2023). Three other key concerns were also raised, namely (1) that the fragmentation of Germany's 'Nature Protected Areas' (there, designated as a formal strict conservation category of land use) leads to extensive contact areas with the surrounding landscape, facilitating spill-over effects from other land uses; (2) agricultural activities such as pesticide/fertiliser applications on adjacent arable lands may harm biodiversity within the protected area and even cause loss of localised plant species near the patch edges; and (3) lack of communication and cooperation between relevant stakeholders prevents optimal insect conservation in the designated protected areas. As elsewhere, stakeholder awareness and support is a vital need for developing policy and reducing threats—and extends to all aspects of insect conservation.

Threats to insects are widespread in protected areas and vary widely across regions. For listed threatened species (of all taxa) in Australia, major management is needed in protected areas to both sustain and restore habitats and control continuing threats (Kearney et al. 2020, 2023), with managing invasive species perhaps the most important of these. The predominant threats differ somewhat in protected areas across the world—fire in Australia may gain more attention than in many other places—but almost universally species in protected areas face multiple threats that require a variety of mitigation actions and which must be integrated with wider management of the area. With only fragmentary knowledge of most insects, management is likely to be generalised—except when focused on individual target species. Even generalised threat management may reduce the proportion of insect species harmed (Chowdhury et al. 2023a, 2023b). Management to increase floristic diversity and heterogeneity, for example, may promote insect diversity, as may increasing availability of waterbodies. Measures such as halting or slowing rates of habitat loss, preventing over-exploitation, and control of alien invasive species become possible within protected areas, with wide benefits to native insects. These are enhanced when management is integrated across the wider landscape.

Four general steps to increasing the values of protected areas for insect conservation, discussed by Chowdhury et al. (2023a), are (1) specifically integrating insects into management plans for existing protected areas, in conjunction with resource preservation and enhancement; (2) strategically designating new protected areas for insects to maximise chances of threat abatement and of conserving metapopulation structures through spatial placement; (3) designing wider insect conservation initiatives beyond the boundaries of protected areas, such as buffer zones, small green spaces (including urban green spaces), and areas with high topographic and hydrological diversity; and (4) investing in insect monitoring and research to improve knowledge of insect diversity and distributions, wherever possible enlisting the help of citizen science (p. 147). The strategy of expanding the protected area network to further embrace coverage of invertebrates is widespread in the fundamental principle and hope of conserving 'habitat' for the greatest variety of species (Morris and Welch 2023a, 2023b). However, a 'watching brief' on the change or expansion of human intrusions into all 'protected areas' may be needed; even small changes such as road widening or stream maintenance may harm critical resources needed by specialised insects. In Australia, mining exploration activities are a continuing concern.

In part, habitat conservation in heavily altered landscapes such as agricultural and urbanised areas needs any formal protected areas to be augmented by attention to remnant vegetation or waterbodies that still may retain some of the otherwise displaced fauna. Small patches of undisturbed vegetation or isolated ponds may be significant refuges, and emulated by more artificial features. Constructed farm dams in South Africa, for example, were regarded as each functioning as a small nature reserve for dragonflies, for which some species were not found in natural habitats in the area (Samways 1989). Areas such as the 'roughs' of golf courses (some of them with ponds), road and rail reserves, and machinery turning areas at field ends and many similar open spaces can sustain native insects through harbouring native vegetation and in addition can be important stepping stones between larger suitable areas or protected areas. In and near urban areas, large vegetated roundabouts and road medians and margins are also significant for insect life both as habitat and in fostering connectivity. Many such areas have not been explored sufficiently to confirm their conservation values, but sensitive management to preserve or restore their natural features is increasingly recommended. The management treatment of such areas may need careful consideration to reduce adverse impacts; on roadside verges and other grassed areas 'arthropod-friendly mowing' with a 'flushing bar' preceding the cutting/mulching mower may have considerable benefits for parasitoid Hymenoptera (Haas-Renninger et al. 2023).

Whilst it may not be known what, or how many, insect species live in a designated 'protected area' it is highly likely that without such areas, many more insects would either have already been lost or become more vulnerable within the local landscape.

For most insect faunas, species representation in protected areas is unknown, so that their real values for insect conservation are also unclear. Such areas are sometimes referred to as a 'last refuge' for species progressively lost from the wider landscape, simply because they are enclaves of less disturbed environments. Some insects are known only from single protected areas: the entire range of the Australian satyrine butterfly *Oreixenica latialis theddora* is within the Mt. Buffalo National Park of alpine Victoria, for example, and some endemic alpine stoneflies (*Thaumatoperla* spp., p. 148) are restricted to high elevation alpine areas, largely within the Alpine National Park.

In contrast, many insects are known only from outside any protected area, in some apparently realistically but for others perhaps reflecting inadequate exploration and survey. In general, narrow-range endemic insects with small distributions may be entirely within protected areas (as in the butterfly above) or not represented. Conversely, many insects are <u>not</u> known from any protected area and, whilst this may sometimes be due to inadequate survey, for many taxa it is probably real. As Chowdhury et al. (2023b) noted, no global study has assessed this representation, and their figure that three-quarters of insect species are insufficiently represented by protected areas leaves little room for complacency. This tentative figure was based on records of the 89,151 insect species with at least three records in the Global Biodiversity Information Facility, and comparing their coverage by protected areas. Of the 827 insect families represented, 525 had at least 75% of the species reported missing a calculated representation target.

Effectiveness of protected areas has been queried from studies enumerating insect group richness within and beyond them—so that the Brazil Conservation Unit's network was not deemed effective for freshwater insects (aquatic Hemiptera: Dias-Silva et al. 2021), in part perhaps reflecting the non-protection of rare species, and that reserves did not encompass species' distributions effectively. The approach of assembling incidence and distributional information on insect groups and examining how 'hotspots' overlap with protected areas has the potential to reveal gaps in coverage, and to imply where additional protected areas may be useful, even if it proves impracticable to pursue. As noted by Kearney et al. (2021), land tenure is a critical influence on what threats may arise but 'The non-protected landscape is a complex matrix of ownership and use regimes' with some not compatible with conservation ideals.

The seven-point 'way forward' proposed for insect conservation in Britain by Morris and Welch (2023a, 2023b) (Table 7.6) relies heavily on information from protected areas, and could be applied elsewhere. The points emphasise the perspective from a reasonably well-known fauna attended by numerous interested

Table 7.6 The 'seven-point plan' for a structured approach to invertebrate conservation in Great Britain, intended to improve identification of important insect sites, site management, and landscape-scale conservation measures (abbreviated from Morris and Welch 2023a, 2023b, used under Creative Commons Attribution licence CC-BY-4.0)

1. Assessment of the available capacity for delivering a robust data set for existing protected areas and data for sites that should be considered as additions to the protected area network
2. Development of a clearer understanding of the microhabitat requirements of threatened species so that site managers recognise the biotopes that support threatened species assemblages
3. Development of better guidance for conservation professionals on the microhabitat requirements of invertebrates to give them higher profile in conservation initiatives such as rewilding and landscape-scale conservation enhancement
4. Analysis of existing data to determine where threatened species occur within the protected sites network. Consider too, whether or not the majority of records are from protected sites or the species involved are effectively 'wider countryside' species for which site protection may be ineffective
5. Evaluation of the temporal structure of the data to establish whether or not there has been an ongoing history of recording at protected sites and if there is a sound knowledge of the likely occurrence of threatened invertebrates
6. Evaluation of the intensity of recording from protected sites so that the records that exist can be contextualised: Do very few records come from a lot of recording effort, or do they come from an equally limited level of surveillance?
7. Expand the protected sites network to reflect deficiencies in coverage of rare and threatened invertebrates and to ensure adequacy of population conservation

naturalists, but the themes they reveal are a sobering reminder of what still remains to be understood. The very limited, and generally non-systematically acquired, documentation of threatened insects in Australia's protected areas largely precludes pursuing the details in this 'wish list' devised for a data-rich fauna. Even for Britain, Morris and Welch commented that 'a significant proportion of … invertebrate fauna is studied by a tiny number of specialists'. However, the central message from this proposal is universal—the key need is to better understand the ecology and microhabitat requirements of insects and then to protect sufficient suitable habitat from damage or loss to ensure that viable populations are sustained. Protected areas are a key element of that approach.

7.3.2 Wider Perspective and Management

The wider landscape of urban and agricultural areas and the multi-use matrix in which these occur presents many opportunities—and demonstrated needs—for insect conservation. The open spaces, including residential spaces, of towns and cities are both extensive and amenable to management. In the United States, for example, Daniels (2013) noted that 'domestic gardens cover millions of hectares' and 'more than 100 million households …have a yard or garden'. 'Butterfly gardening', the deliberate plantings of nectar plants and larval food plants in home gardens and accompanied by reduced pesticide uses, is popular in many places, and its wider benefits include encouragement of other insects seeking similar resources whilst retaining aesthetic appeal. Advocacy for garden plantings to deliberately encourage pollinators extends the above needs to include 'bee hotels' (valuable also as educational aids), dead wood, and other nesting sites as additional critical resources, although some reluctance to deploy old stumps and brush cuttings may occur, as countering the traditional wish for 'neatness' in home and public gardens. Pollinators respond positively to features such as heightened plant diversity, woody vegetation, garden size, and sun exposure, but negatively to garden isolation from natural areas (Majewska and Altizer 2018). Habitat restoration in urban areas involves enrichment and informed treatment, with needs to promote connectivity wherever possible. However, gardens may not wholly reflect the capability of more natural environments to sustain pollinators, their natural enemies, and other insects. Maximising their potential demands greater understanding of what factors are beneficial or detrimental to the insects (Majewska and Altizer 2018). With focused management, urban gardens can promote foraging activities of stingless bees to reach levels higher than in nearby forest/plantation habitats (Australia: Kalusa et al. 2016). As Baldock (2020) concluded, the numerous opportunities for pollinator conservation in urban areas reflect activities of residents, local authority, policymakers, urban planners, and landscape architects—the last two influencing the design and ecological features of new and restored developments. The five main principles of 'biodiversity-sensitive urban design' (Garrard et al. 2018) have wide relevance to insects, as (1) maintaining and introducing habitat; (2) facilitating dispersal; (3) minimising threats and anthropogenic disturbances; (4) facilitating natural ecological processes; and (5) improving potential for positive human-nature interactions.

The changing attitude to lawn management to increase conservation value exemplifies wider trends for urban spaces. Lawn areas commonly sum to comprise more than half the area of urban open spaces. The cultural or 'traditional' approach to urban grass lawn management has emphasised neatness and uniformity, with well-kept turfgrass lawns 'associated with good moral character, neighbourliness, and higher property values' (Ramer et al. 2019), with appearance of lawns often considered their most important feature! Many are based on monocultures of particular grass varieties, and are mown and fertilised/watered regularly; many also need at least moderately intensive management such as mowing and weeding. However, as advocated for pollinator conservation, incorporating low-growing flowers into mown lawns can provide important additional resources whilst also maintaining traditional uses and general appearance. Going further, alternatives to turfgrass lawns can include 'grass-free lawns' of perennial forbs or (mainly in public areas) 'urban meadows' with higher biodiversity benefits.

Many urban open spaces are small and isolated. Promoting functional linkages, such as corridors (below), is increasingly important in urban design, perhaps especially in aiding connectivity between remaining natural vegetation patches. More broadly, and extending from urban areas alone, roadside environments are an undervalued component of insect habitats, and far more extensive than generally supposed. In Australia roadside strips in rural areas may support the only remaining natural vegetation in highly transformed agricultural landscapes and may themselves become vulnerable to activities such as road widening and realignment. Their values as insect habitats and refugia, together with their values for plant conservation, are increasingly recognised (New et al. 2021). Phillips et al. (2020) quoted an earlier figure of 36 million Km of road length across the world, and other data on the areas of road verges endorse their values as 'habitat' and for connectivity: Phillips et al. noted estimates of 2400 Km^2 for Britain and 50,000 Km^2 for the United States, representing 1% and 0.5% of land areas, respectively.

Insect conservation in agroecosystems has a rather different emphasis, through needs to integrate this over large areas in which the primary aim is to protect and enhance their productive capability, and to suppress measures—such as pests—that threaten or reduce this in ways that do not reduce market value. Pest management must be undertaken in the most economical, effective, and environmentally safe ways possible. The early dependence on broad-spectrum chemical pesticides and the environmental concerns these raised from the mid-twentieth century onwards stimulated development of varied components of 'integrated pest management' whereby biological and other non-polluting measures form the core of management, and pesticide applications are reduced to essential targeted levels and with minimal undesirable side effects. The wider needs of crop production include pollination, protection of all growth stages from pest attack reaching economically damaging levels, and any later post-harvest storage. However, the highly nutritious cultivars of many food crops are attractive to insect consumers. Insect conservation in such artificial crop or pasture environments draws on measures to encourage pest suppression by naturally occurring predators and parasitoids (together, with pathogens 'natural enemies'), and the environmental modifications needed to promote their activity, broadly as 'conservation biological control' (p. 43). Those beneficial insects need (1) habitat space as refuges from disturbances, for continuing to thrive over periods when the crop is not present, for overwintering and general sustenance, and (2) reliable food supply for periods when the pest is not available and for developmental stages not associated with the pests. A variety of natural and 'semi-natural habitats' in agricultural areas could include remnant patches of native vegetation, hedgerows, fallow areas, and field margins, all amenable to management to promote desirable insects and of far wider value for others—but also with the perception from farmers that they can also harbour reservoirs of pest insects. Floral enrichment of field margins and additional constructions such as 'beetle banks' add to both habitat availability and connectivity. Beetle banks are raised rows of grasses and associated perennial vegetation constructed as within-field non-cropping areas to extend into fields from field edges, to promote in-crop activity by, in the original focus, predatory beetles such as Carabidae in Europe (Thomas et al. 1991), and which may serve as refuges for natural enemies for a decade or more (Thomas et al. 2002). Natural enemies may respond positively to management measures such as increased plant variety (including larval foods and nectar/pollen supply for the corresponding adults), enhanced landscape complexity and heterogeneity, and reduced cropping intensity. Much conservation biological control is focused wholly on agricultural areas, but the principles extend far wider, to encouragement of natural enemies over a landscape scale, within which crop protection needs are embedded. Thus, the 'Greening Waipara' project in the wine-growing areas of Canterbury, New Zealand, sought to restore ecological services and functional biodiversity to agricultural ecosystems by establishing native plant groundcovers to control weeds in vineyards, and wider plantings within the region. Inter-row plantings of flowering plants increased the availability of parasitoids to control pest leafroller moths, and the wider plantings of native flora in the Waipara Valley fostered wider conservation and ecotourism (Forbes et al. 2013). The project also increased the values of vineyards and their surrounds for butterflies (Gillespie and Wratten 2012), although these were most dependent on remnant vegetation patches.

That project also demonstrated the now widespread importance of retaining natural or seminatural vegetation on farmlands and of restoration to expand this for biodiversity conservation. These may not necessarily be extensive; retention of scattered paddock trees, as stock shelter for example, can provide both habitat and connectivity for insects. Restoration can range from 'leave alone' to facilitate natural regeneration to more active focus such as plantings of native or other vegetation, perhaps with roles such as shelterbelts or woodlots viewed positively by the landowners. Native and restored vegetation on farms in temperate Australia contributes to biodiversity conservation, agricultural production, and human wellbeing, with retention of native vegetation the most important beneficial factor (Haslem et al. 2023).

Insect diversity conservation in any landscape devolves firmly on habitat extent, condition, and connectivity, the last—in particular—incorporating the landscape in which protected areas and agricultural and urban transformations are embedded. Projections to improve 'connectivity' through (1) establishing corridors or 'stepping stones' for insect movement and (2) reducing or removing barriers are common in insect conservation plans, for both individual species and broader faunas. However, the concepts around these purported linear habitats are rather more complex than often supposed (Dennis et al. 2013), as are evaluations of 'fragmentation' based on distance between habitat patches (Hanski and Poyry 2007), and such plans need careful consideration. Dennis et al. suggested, from considering butterflies, that specialists and generalists have different needs for connectivity. Both benefit from larger habitat patches, but specialists need short-range corridors and generalists can use stepping stones over larger distances. Linkages between occupied patches and others that are also occupied or are potentially habitable have several functions—reducing probability of local extinction, facilitating colonisation or recolonisation, and enabling changes in range in response to climate changes. The population structure of many insect species remains unknown—but any form of metapopulation structure—whereby continuing colonisations and extirpations across multiple habitat patches in a region sustain the overall population in the landscape—also depends on continuing accessibility of those patches. Connectivity, and facilitating insect dispersal, requires some network of effective movement, and conservation planning may consider this constructively.

Conventionally, and often without direct evidence, particular structural landscape features are categorised as either 'corridors' or 'barriers' affecting insect dispersal. In particular, linear features such as hedgerows, field margins, roadsides, green lanes, powerline and rail easements, riparian strips, and others are presumed to facilitate and enable movements of species, whereas structures such as motorways, rail lines, agricultural fields, and alien conifer plantations are regarded as barriers in impeding movements. The great variations in insect dispersal capability provide an almost infinite range of possibilities and behaviour, so that dispersal (and the factors influencing it) differs across individuals, species, ages, sexes, and influences of the local structural environment and the resources present and climate. Nevertheless, much of the conventional planning for insect conservation focuses on habitat size, condition, and connectivity through linear landscape features. Those presumed to be corridors, providing potential linkages between patches of suitable or remnant habitats, may also support resident populations as well as being conduits, with this role encouraged by wide corridors in which edge effects are minimal. Samways (2007a, b) and others have emphasised the importance for insects of establishing regional networks of corridors, not least to counter isolation from impacts of gradual climate change. It may be feasible to establish this principle to accompany anthropogenic changes in landscapes. In South Africa, development of agroforestry has involved large-scale plantation establishment of conifers. The forestry industry responded to criticisms of early impacts on native biodiversity, by leaving a network of remnant grassland linkages and nodes between plantation stands, with the total land sparing being about a third of the land area, in addition to proclaimed reserves (Samways 2007b). Wide linkages, more than 200 m wide, were adopted as habitat by several butterflies, some penetrating deeply into the network from more distant native grasslands. Disturbance of the corridors by domestic cattle led to changes in floral diversity and structure, adversely affecting insects, but this development led to design guidelines for such ecological networks (Samways 2007b).

An ambitious UK project initiated by Buglife, the Invertebrate Conservation Trust, draws on widespread concerns over pollinator declines and involves plans for a nationwide network of wildflower floristically enriched pollinator 'superhighways' that can act as stepping stones or corridors to facilitate pollinator movement. The concept was developed as part of the response to the 'National Pollinator Strategy' adopted in 2014, and was the subject of a Private Member's Bill introduced to the UK Parliament House of Commons in May 2018 (Bradley 2018). 'B-lines' is a network of 3 Km-wide pathways connecting the best remaining wildflower habitats and designed to reconnect these landscapes across the entire United Kingdom (Buglife 2021). The network was designed through activities of numerous partners, at regional or county level, mapping the best wildflower-rich habitats and indicating potential connection routes between them and with local authorities and community groups helping to finalise the mapped networks. Buglife then works with local groups and experts so that local knowledge can guide developments, and ensure the communities are invested in the development of 'their' area. An interactive national map shows the very best connection routes refined by computer modelling of the initial plans, and can be used by others to see where activities are needed and to monitor progress. Some routes reflect existing corridors such as footpaths, rights-of-way, and roadsides. Others may span the regions between existing protected areas or important sites, such as areas with significant plant diversity and notable flowering species.

Many roadsides in Australia are signed individually as supporting 'significant flora', the signs enabling maintenance workers, such as those slashing or mowing road verges, to avoid these: the 'long paddock' (travelling stock routes: TSRs) used for moving stock along Australian roadsides may contain significant proportions of the remaining native vegetation and provide refuges for threatened plant species; TSRs have also been claimed to be critically important for conservation of fauna as providing a continuous network of habitat corridors. Many TSRs have high conservation status (Spooner et al. 2010), in conjunction with significant cultural and historical values through being founded on traditional aboriginal pathways (used long before European settlement) and routes established by early explorers. Some were founded on water availability, with the routes linking water sources at intervals of 'a day's walk'. More recently, some TSRs have marked the position of major traffic highways.

In Britain, many community groups have adopted sites along the planned B-lines for plantings of local wildflowers and wider rehabilitation—measures that will benefit not only pollinators but a host of other native insects that participate in related food webs. The scheme has recently been adopted in the European Union, with ecological corridors to promote pollinator movements proposed to be termed 'Buzz lines' (European Commission 2023).

Evaluating the success of any such scheme (as with many other management steps) depends on effective monitoring—repeated inspections or sampling to detect incidence and abundance of the pollinators and changes over time. Although with slightly different emphasis, Breeze et al. (2020) noted that 'long-term monitoring of pollinators is urgently needed' to provide consistent baseline information against which to measure changes, and considered the costs of doing this in Britain by using several different approaches to contribute to two major interest groups. These are (1) research workers, by designing the network to address scientific questions, and (2) society, by offsetting the risks from failing to respond to pollinator declines that affect biodiversity, crop pollination services, and human wellbeing. Breeze et al. contrasted the costs of monitoring by professionals and volunteers, using (hypothetical) defined sampling protocols and evaluating the contributions possible. The higher costs of 'professionals' were associated with more rigorous quantitative data and species-level identifications, compared with less penetrating identifications (as 'morphospecies') by volunteer observations from pan trap catches and direct focal examinations of flower visitors. The limited support likely to be available for any such exercise demands very careful consideration of the level of information needed. Whilst it is usually straightforward to classify pollinators or flower visitors into major groups as 'hoverflies', 'bees', 'wasps', 'beetles', 'moths', and so on, species-level naming of any of these groups is far more complex and laborious, and usually demands expert participation.

The need to establish ‘baselines’ of insect richness and abundance against which future changes can be monitored must be focused on particular insect groups such as those noted above. The parameters for selecting those focal groups, as listed by Lewinsohn et al. (2022) for Brazil—also with very incompletely documented insect richness and distributions—deserve consideration in setting priorities for Australia, and are (1) groups with well-founded taxonomy and accessible in-country reference collections and specialists; (2) groups clearly associated with particular ecosystem functions or processes; and (3) groups that can be recorded or collected by non-specialists in citizen science programmes.

References

Ahern LD, Tsyrlin E, Myers R (2003) Mount Donna Buang wingless stonefly *Riekoperla darlingtoni*, Action Statement no 125. Department of Sustainability and Environment, Victoria

Allan PBM (1943) Talking of moths. Newtown, Montgomery

Baldock KCR (2020) Opportunities and threats for pollinator conservation in global towns and cities. Curr Opin Insect Sci 38:63–71

Balzani P, Dekoninck W, Feldhaar H, Freitag A, Frizzi F (et al.) (2022) Challenges and a call to action for protecting European red wood ants. Conserv Biol 36 (6) e13959 https://doi.org/10.1111/cobi.13959

Bellis J, Bourke D, Williams C, Dalrymple S (2019) Identifying factors associated with the success and failure of terrestrial insect translocations. Biol Conserv 296:29–36. https://doi.org/10.1016/j.biocon.2019.05.008

Boermsma PD, Kareiva P, Fagan WF, Clark JA, Hoekstra JM (2001) How good are endangered species recovery plans? Bioscience 51:643–649

Bolam FC, Ahumada J, Akcakaya HR, Brooks TM, Elliott W (et al.) (2022) Over half of threatened species require targeted recovery actions to avert human-induced extinction. Front Ecol Environ 21: 64–70. https://doi.org/10.1002/fee.2537

Braby MF (2018) Threatened species conservation of invertebrates in Australia: an overview. Aust Entomol 57:173–181. https://doi.org/10.1111/aen.12324

Bradley B (2018) Private members bill: protection of pollinators. UK Parliament Hansard 640: debated 8 May 2018

Breeze TD, Bailey AP, Balcombe KG, Breteon T, Comont R (et al.) (2020) Pollinator monitoring more than pays for itself. J Appl Ecol 58: 44–57. https://doi.org/10.1111/1365-2664.13755

Buglife (2021) B-lines. Insect superhighways Buglife: the invertebrate conservation trust, Peterborough

Chowdhury S, Dubey VK, Choudhury S, Das A, Jeengar D (et al.) (2023a) Insects as bioindicator: a hidden gem for environmental monitoring. Front Environ Sci 11:11.46052. https://doi.org/10.3389/fenvs.2923.1146052

Chowdhury S, Zalucki MP, Hanson JG, Tiatragul S, Green D (et al.) (2023b) Three-quarters of insect species are insufficiently represented by protected areas. One Earth 6: 139–146. https://doi.org/10.1016/j.oneear.2022.12.003

Clausnitzer V, Kalkman VJ, Ram R, Collen B, Baillie JEM (et al.) (2009) Odonata enter the biodiversity crisis debate: the first global assessment of an insect group. Biol Conserv 142: 1864–1869. https://doi.org/10.1016/j.bicon.2009.03.028

Collins NM (1987) Legislation to conserve insects in Europe. Pamphlet 13,. Amateur Entomologist’s Society

Daniels J (2013) Gardening and landscape modification: butterfly gardens. In: Lemelin RH (ed) The management of insects in recreation and tourism. Cambridge University Press, Cambridge, pp 153–168
DEH (Department of Environment and Heritage) (2007) Advice to the Minister for the Environment and Heritage from the Threatened Species Scientific Committee – *Synemon discalis* (Small Orange-spotted Sun-moth). http://www.enviroment.gov.au/biodiversity/threatened/species/synemon-discalis.html
Dennis RLH, Dapporto L, Dover JW, Shreeve TG (2013) Corridors and barriers in biodiversity conservation: a novel resource-based habitat perspective for butterflies. Biodivers Conserv 22:2709–2734. https://doi.org/10.1007/s10531-013-0540-2
Dias-Silva K, Vieira TB, Moreira FFF, Juen L, Hamada N (2021) Protected areas are not effective for the conservation of freshwater insects in Brazil. Sci Rep 11:21247. https://doi.org/10.1038/s41598-021-00700-0
Dolny A, Sigutova H, Ozana S, Choleva L (2018) How difficult is it to reintroduce a dragonfly? Fifteen years monitoring *Leucorrhinia dubia* at the receiving site. Biol Conserv 218:110–117. https://doi.org/10.1016/j.biocon.2017.12.011
Douglas F (2003) Five threatened Victorian sun-moths (*Synemon* species). Flora and Fauna Guarantee Act 1988. Action Statement no. 146. Department of Sustainability and Environment, Victoria
Duffus NE, Morimoto J (2022) Current conservation policies in the UK and Ireland overlook endangered insects and are taxonomically biased towards Lepidoptera. Biol Conserv 266:109464. https://doi.org/10.1016/j.biocon.2022.109464
European Commission (2023) Communication for the commission to the European Parliament, the council, the European economic and social committee and the Committee of the Regions. Revision of the EU Pollinators Initiative A new deal for pollinators. COM/2023/35 final
Forbes SL, Cullen R, Grout R (2013) Adoption of environmental innovations: analysis from the Waipara wine industry. Wine Econ Policy 2:11–18
Garrard GE, Williams NSG, Mata L, Thomas J, Bekessy SA (2018) Biodiversity sensitive urban design. Conserv Lett 11:1–10. https://doi.org/10.1111/conl.12411
Gibson L, New TR (2007) Problems in studying populations of the golden sun-moth, *Synemon plana* (Lepidoptera: Castniidae), in south eastern Australia. J Insect Conserv 11:309–313
Gillespie M, Wratten SD (2012) The importance of viticultural landscape features and ecosystem service enhancement for native butterflies in New Zealand vineyards. J Insect Conserv 16:13–23. https://doi.org/10.1007/s10841-011-9390-y
Haas-Renninger M, Weber J, Felske I, Kimmich T, Csader M (et al.) (2023) Microhymenoptera in roadside verges and the potential of arthropod-friendly mowing for their preservation. J Appl Entomol 147.10: 1035–1044. https://doi.org/10.1111/jen.13199
Hallmann CA, Sorg M, Jongejans E, Siepel H, Hofland N (et al.) (2017) More than 75 percent decline over 27 years in total flying insect biomass in protected areas. PLoS One 12; e0185809.https://doi.org/10.1371/journal.pone.0185809
Hanski I, Poyry J (2007) Insect populations in fragmented habitats. In: Stewart AJA, New TR, Lewis OT (eds) Insect conservation biology. CAB International, Wallingford, pp 175–202
Harvey JA, Heinen R, Armbrecht I, Basset Y, Baxter-Gilbert JH (et al.) (2020) International scientists formulate a roadmap for insect conservation and recovery. Nat Ecol Evol 4: 174–176
Haslem A, Bennett AF, Radford JQ (2023) Importance of (semi)natural vegetation on farms for achieving multiple objectives: a conceptual model based on temperate southern Australia. Agric Syst 213:103795. https://doi.org/10.1016/j.agsy.2023.103795
Hochkirch, A, Nieto A, Criado MG, Calix M, Braud Y (et al.) (2016) European red list of grasshoppers, crickets and bush-crickets. Publications Office of the European Union, Luxembourg
Honan P (2008) Notes on the biology, captive management and conservation status of the Lord Howe Island stick insect (*Dryococelus australis*) (Phasmatodea). J Insect Conserv 12:399–413
IUCN (2012) IUCN red list categories and criteria: version 3.1, 2nd edn. IUCN, Gland

IUCN (2013) IUCN guidelines for reintroduction and other conservation translocations. IUCN Species Survival Commission, Gland
IUCN (2017) Guidelines for using the IUCN red list categories and criteria, Version 13. IUCN, Gland
Kalusa BF, Wallace H, Heard TA, Klein A-M, Leonhardt SD (2016) Urban gardens promote bee foraging over natural habitats and plantations. Ecol Evol 6:1304–1316
Kearney MR, Hossain MA, Sinclair SJ, Song H (2020) Grasshopper country before and after: a resurvey of Ken Key's collecting expeditions in New South Wales, Australia, 70 years on. Austral Entomol 60:52–65. https://doi.org/10.1111/aen.12515
Kearney SG, Carwardine J, Reside A, Adams VM, Nelson R (2021) Saving species beyond the protected area fence: threats must be managed across multiple land tenure types to secure Australia's endangered species. Conserv Sci Pract 4:e617. https://doi.org/10.1111/csp2.617
Kearney SG, Watson JEM, Reside AE, Fisher DO, Maron M (et al.) (2023) Threat-abatement framework confirms habitat retention and invasive species management are critical to conserve Australia's threatened species. Biol Conserv 277: 1–11. https://doi.org/10.1016/j.biocon.2022.109833
Kothe S, Bakanov N, Bruhl CA, Eichler L, Fickel T (et al.) (2023) Recommendations for effective insect conservation in nature protected areas based on a transdisciplinary project in Germany. Environ Sci Europe 35: 102. https://doi.org/10.1186/s12302-023-008135
Leandro C, Jay-Robert P, Vergnes A (2017) Bias and perspectives in insect conservation: a European scale analysis. Biol Conserv 215:213–224. https://doi.org/10.1016/j.biocon.2017.07.033
Lewinsohn TM, Agostini K, Freitas AVL, Melo SAS (2022) Insect decline in Brazil: an appraisal of current evidence. Biol Lett 18:20220219. https://doi.org/10.1098/rsbl.2022.0219
Lewis SM, Jusoh WEA, Walker AC, Fallon CE, Joyce R, Yiu V (2024) Illuminating firefly diversity: trends, threat and conservation strategies. Insects 15:71. https://doi.org/10.3390/insects15010071
Lindenmayer D, Woinarski J, Legge S, Southwell D, Lavery T (et al.) (2020) A checklist of attributes for effective monitoring of threatened species and threatened ecosystems. J Environ Manag 262 (2020) 110312. https://doi.org/10.1016/j.jenvman.2020.110312
Majewska AA, Altizer S (2018) Planting gardens to support pollinators. Conserv Biol 34(1):15–25. https://doi.org/10.1111/cobi.13271
Mjadwesch R, Nally S (2008) Emergency relocation of a purple copper butterfly colony during roadworks: successes and lessons learned. Ecol Manage Restor 9:100–109. https://doi.org/10.1111/j.1442-89803.2008.00400.x
Morris R, Welch M (2023a) Institutional vertebratism is alive (and kicking invertebrates). Antenna 47:23–27
Morris RKA, Welch MD (2023b) Is invertebrate conservation in Great Britain best achieved by policies that increase species protection? J Insect Conserv 27:527–531. https://doi.org/10.1007/s10841-023-00485-9
Nason SE, Lloyd N, Kelly CD, Brichieri-Colombi T, Dalrymple SE, Moehrenschlager A (2021) Maximising the effectiveness of quantitative systematic reviews: a case study on terrestrial arthropod conservation translocations. Biol Conserv 254:108948. https://doi.org/10.1016/j.biocon.2020.108948
New TR (2009) Insect species conservation. Cambridge University Press, Cambridge
New TR, Yen AL (1995) Ecological importance and invertebrate conservation. Oryx 29:187–191. https://doi.org/10.1017/50030605300021116
New TR, Yen AL (2013) Invertebrate conservation in Australia: problems in policy and practice. Pac Conserv Biol 19:104–109. https://doi.org/10.1071/PC130104
New TR, Sands DPA, Taylor GS (2021) Roles of roadside vegetation in insect conservation in Australia. Aust Entomol 60:128–137
Pearce-Kelly P, Ferguson A, McFarlane D, Pearce K, Sainsbury AW (et al.) (2024) Breeding and relocating insects for conservation. pp. 197–209 in Pryke JS, Samways MJ, New TR,

Cardoso P, Gaigher R (eds) Routledge handbook of insect conservation. Routledge, London and New York

Phillips BB, Wallace C, Roberts BR, Whitehouse AT, Gaston KJ (et al.) (2020) Enhancing road verges to aid pollinator conservation: a review. Biol Conserv 250: 108687 https://doi.org/10.1016/j.biocon.2020.108687

Rada S, Schweiger O, Harpke A, Kuhn E, Kuraz T (et al.) (2019) Protected areas do not mitigate biodiversity declines: a case study on butterflies. Divers Distrib 25: 217–224. https://doi.org/10.1111/ddi.12854

Ramer H, Nelson KC, Spivak M, Wolfin J, Pulscher ML (2019) Exploring park visitor perceptions of 'flowering bee lawns' in neighborhood parks in Minneapolis, MN, US. Landsc Urban Plan 189:117–128. https://doi.org/10.1016/j.landurbplan.2019.04.015

Samways MJ (1989) Farm dams as nature reserves for dragonflies (Odonata) at various altitudes in the Natal Drakensberg Mountains, South Africa. Biol Conserv 48:181–187

Samways MJ (2007a) Insect conservation: a synthetic management approach. Annu Rev Entomol 52:465–487. https://doi.org/10.1146/annurev.ento.52110405.091317

Samways MJ (2007b) Implementing ecological networks for conserving insect and other biodiversity. In: Stewart AJA, New TR, Lewis OT (eds) Insect conservation biology. CAB International, Wallingford, pp 127–143

Samways MJ (2020) Insect conservation a global synthesis. CAB International, Wallingford

Samways MJ, McGeoch MA, New TR (2010) Insect conservation. A handbook of approaches and methods. Oxford University Press, Oxford

Sands DPA, New TR (2002) The action plan for Australian butterflies. Environment Australia, Canberra

Sands DPA, New TR (2003) Coordinated invertebrate surveys in Australia's national parks: an important tool in refining invertebrate conservation management. Rec S A Mus, Monogr series 7:203–207

Sherley GH (1995) Threatened weta recovery plan. Threatened species recovery plan no 25. Department of Conservation, Wellington

Spooner PG, Firman M, Yalmambirra (2010) Origins of traveling stock routes.1. Connections to indigenous traditional pathways. Rangel J 32:329–339. https://doi.org/10.1071/RJ10009. 10.1036-9872/10/030329

Stubbs A, Shardlow M (2012) The development of Buglife—the Invertebrate Conservation Trust. In: New TR (ed) Insect conservation: past, present and prospects. Springer, Dordrecht, pp 75–106

Taylor GS, Braby MF, Moir ML, Harvey MS, Sands DPA (et al.) (2018) Strategic national approach for improving the conservation management of insects and allied invertebrates in Australia. Aust Entomol 57: 124–149. https://doi.org/10.1111/aen.12343

Thomas MB, Wratten SD, Sotherton NW (1991) Creation of 'Island' habitats in farmland to manipulate populations of beneficial arthropods: predator densities and emigration. J Appl Ecol 28:906–917

Thomas SR, Goulson D, Holland JM (2002) The contribution of beetle banks to biodiversity. Asp Appl Biol 62:31–38

Wagner DL (2020) Insect declines in the Anthropocene. Annu Rev Entomol 65:457–480. https://doi.org/10.1146/annurev-ento-011019-025151

Woinarski JCZ, Chapple DG, Garnett ST, Legge SM, Lintermans M, Scheele BC (2023) Few havens for threatened Australian animal taxa that are highly susceptible to introduced and problematical native species. Biodivers Conserv 33(1):305–331. https://doi.org/10.1007/s10531-023-02750-4

Yagui H, Kearney MR, Hoffmann AA (2024) Restoring declining species through translocations: a test case using flightless grasshoppers in an urban setting. Insect Conserv Divers 17:358–373. https://doi.org/10.1111/icad.12679

Chapter 8
Involving the Wider Community

8.1 Introduction

Insect conservation is a massive task! Its needs transcend hindrances from prejudice and ignorance and necessitate the goodwill and effective support amongst policy-makers, scientists, conservation organisations and managers, young people, and citizens across all walks of life—to many of whom those needs may not be apparent or easily accepted and to whom, also, insects may not be accepted easily as worthy of conservation. Education and effective communication are central channels to gaining public and political sympathy and support. This task is itself daunting, and the limited logistic support likely to be available for insect conservation activities—including finance and expertise—necessitates very clear focus of purpose and priorities. And, although their numbers and influence are increasing, 'insect conservationists' themselves are a small constituency! Some conservation agencies have no dedicated entomologists on permanent staff, so that their remit to 'conserve biodiversity' then lacks a key focal area of expertise. Many entomological and broader natural history societies have a conservation committee or members interested in conservation, and broader interest is evident from now long-established groups such as the Xerces Society for Invertebrate Conservation (North America) and Butterfly Conservation (initially British but now Butterfly Conservation Europe) and Buglife—The Invertebrate Conservation Trust, which have spear-headed national and wider interests. Their development is outlined by contributors to New (2012), and their wider influences have stimulated citizen-based groups elsewhere. Largely due to such societal efforts, some awareness of insect significance and diversity has increased substantially, and the consequences of insect losses are far better understood. Interest in insect natural history, the foundation of much conservation assessment and practice, has benefited from progressive recording and survey, to lead to databases, coordinated distribution records, phenological information, and resource/habitat needs as accumulated in forms that can be interpreted for a

T. R. New, *Insect Conservation in Australia: Why and How*,
https://doi.org/10.1007/978-3-031-66631-5_8

variety of purposes. Such associations are effective venues of communication, and needs for information on any nominated species, group, habitat, site (and others) may be solicited or given priority amongst their membership as needs or special interests arise. They are also important sources of information for managers and others seeking entomological advice, or on where else it might be available. The extension of insect conservation interests into the wider community incorporates understanding the human dimensions involved, so invoking aspects of 'conservation social science' (Bennett et al. 2016, 2017). This need reflects the reality that many problems of practical conservation cannot be solved by scientists (and scientific methods) alone, but are largely societal issues. Supply of information and providing the 'evidence base' to the variety of stakeholder interests likely to be involved in any individual case is a critical lever in garnering community support.

8.2 Education

Reduced interest in outdoor pursuits involving 'natural history', accompanied for insects by collecting now having declined as a hobby—and in some places essentially stigmatised as socially and ethically unacceptable—has led to many fewer young people developing an interest in entomology over recent decades. Their possible mentors are also fewer in number, a situation sometimes dubbed 'the extinction of experience' and eroding a long tradition of natural history interest in Britain and some other places (Cheesman and Key 2007). Outdoor learning for young people is far more memorable and creates more lasting impressions than indoor instruction or screen-based learning—school or later 'field trips' are highlights for many students, but are progressively more difficult to undertake as costs and social insecurities increase. Calls for increased inclusion of entomology in school science curricula continue to suggest numerous innovative opportunities for stimulating interest—but with little consistently positive outcome. Innovative and varied programmes focused on either single species, such as the exercises based on the parasitoid wasp *Melittobia* (designated 'WOWBugs' ['Working On Wasps' or 'Wild Over Wasps': Matthews 2000]), or wider diversity exemplify far wider possibilities (Matthews et al. 1997) for realising development of 'tomorrow's scientists from today's schools'.

Further, and also of considerable concern, formal instruction in traditional entomology in many undergraduate 'biology' courses across the world has also declined widely. Since the 1970s, the number of entomology-specific courses or subjects taught in Australian universities has 'plummeted' (Merritt 2023), with changes in educational emphasis and closures or amalgamations of subjects with low enrolments. Importantly, reductions of (expensive) laboratory practical classes and field trips have reduced student exposure to insect life. Rather than gaining a solid grounding in entomology during their undergraduate courses, many biology students now only gain more selective knowledge whilst employed or undertaking higher degree studies, and commonly lack a 'rounded appreciation' of insect life and diversity.

More generally, coverage of insects in many biology textbooks has also declined: comparisons of the numbers of pages devoted to insects and numbers of illustrations of insects in a wide selection of texts implied that such content had declined by more than 75% in recent (published since 2000) compared with older (published before 1960) texts (Gangwani and Landin 2018). Environmental educational programmes can do much to promote insect conservation to people of all ages (Sitar and Rusu 2023), but are perhaps of particular importance in moulding the attitudes and perceptions of young people. Occupations such as home gardening can help encourage interests in natural history and increase positive feelings towards insects (Vanderstock et al. 2022) and lead to greater appreciation of their variety.

Advancing insect conservation depends critically on the interests and participation of hobbyists and other 'non-professional' entomologists, whose knowledge in many instances by far exceeds that of any formally designated entomological project leader through whom a conservation case may be advanced. They may have long and detailed experience of either particular insect taxa or significant sites, and be amongst the first to become aware of early changes or impending threats. Many of the species listed for conservation need were signalled through advocacy from local naturalists concerned over perceived declines or threats and, reflecting paucity of formal monitoring of the abundance and incidence of most insects, this welcome 'ad hoc' process is likely to continue. Even though it may not alone be an adequate pathway to insect conservation, single-species conservation is a major avenue to understanding and publicising the need through education (including young people), scientific leadership, and citizen science. Collectively, these three themes are continuing keys that go hand in hand in advancing insect conservation. Without community-based support networks, much insect conservation awareness and activity would be impracticable other than by the passive benefits of 'protected areas' such as reserves and national parks, most of which have been established primarily for other purposes and are not managed for promoting insect diversity. Whether for single species or wider diversity, increasing awareness and improving the public's image of insects is key to garnering support. Practical exercises such as establishing B-lines (p. 136) for pollinators could clearly not be pursued without this wider participation, aided by support from local authority and higher-level policy.

'Visibility' of insects can be encouraged in many ways. Venues such as 'butterfly houses' and similar exhibits in which living insects can be viewed remain popular, and are amongst the most visited exhibits in many institutions. Although their primary purpose has historically been 'entertainment', many zoos and related bodies have broadened their interests and remit to include conservation breeding and participation in invertebrate conservation programmes across the wider community. Their accumulated experiences in breeding, effective disease control and sanitation, and wider insect husbandry transfer to more primarily conservation ex situ programmes elsewhere (p. 124). Practical measures such as 'butterfly gardens', 'bee hotels', and preserving or digging ponds for dragonflies are all well-tried and widespread conservation adjuncts that can be constructed and monitored by individuals or community groups. They function by providing suitable conditions and resources, in some cases augmenting those already present. 'Dragonfly ponds' have become a widespread conservation exercise (Moore 1997), with wider benefits as refuges for other aquatic invertebrates. They may, however, require continuing management. Changes from increased vegetation, sediment accumulation, overshading, and bank conditions may all affect the diversity of microhabitats needed to sustain a varied dragonfly assemblage. Some form of 'constructed wetland' as a scenic recreation area is common in new suburbs associated with urban expansion in Australia. They, and many similar exercises, depend on creating environments that can be colonised naturally, or to which the insects can be translocated, where they can be viewed easily by the public, and where informative educational labelling and aids to recognition can introduce insect variety and biology. Notice of any unusual biological habits can stimulate interest. 'Awareness' of insects and their needs links strongly with visibility, and 'popular' or 'iconic' insect groups such as butterflies and dragonflies that are visible during the day can lead to interest in other, less conspicuous, taxa. A walkable 'dragonfly awareness trail' developed in the National Botanic Gardens of Pietermaritzburg (South Africa) included four ecologically different biotopes with stations from which selections of the 35 species of Odonata present in the Gardens could be viewed (Suh and Samways 2001). An information booklet introduced the insects, their identification, and their ecological variety. Changes have been monitored and endorse the needs for management noted above to maintain this diversity. Over 32 years, including comparisons with Odonata present from 1991, before the sampled pond ('Kingfisher Lake', along the awareness trail) was constructed, sustaining dragonflies through habitat condition and variety depended on regular management (Kietska et al. 2021), with variations reflecting decreased conditions through sediment accumulation and vegetation condition at intervals, and increases once remedial management occurred.

In a somewhat different context, exemplifying the beneficial roles of ecological pest management, the 'Biodiversity trail' suggested for selected New Zealand vineyards by the Greening Waipara project (p. 135) was designed to provide winery visitors/customers with a wider experience by leading them through areas of vines and native plants, with informative notices on biodiversity and management by diversification (Forbes et al. 2013).

Other avenues of publicity may be available—a children's playground in outer Melbourne is modelled on the Golden sun-moth (*Synemon plana*, Castniidae), and accompanying illustrated notices (see New 2015) drew public attention to the plight of the moth on nearby periurban grasslands. Likewise, and also in Melbourne, publicity for the Eltham copper butterfly (*Paralucia pyrodiscus lucida*, Lycaenidae) also included a playground in a local park. Many parallels help individual insect conservation targets based on flagship species to gain public acceptance—they are largely 'taken for granted' in countries with a long practical history of interest in insect conservation but are still novel elsewhere, where reluctance to accept insects as 'worthy' persists widely. As with developing and promoting other aspects of conservation for which economic returns are elusive, sound direction and encouragement may prove pivotal.

Throughout the world, many people are introduced to insect conservation through attention drawn to a single nominated and distinctive species in their local area, and need to survey for it, detect and monitor its incidence, and evaluate threats and recovery actions. One of the most fundamental needs is to consistently and accurately recognise the insect(s) involved, an obvious prerequisite to study but not always heeded carefully! A clear diagnosis, listing characteristic recognition features and preferably illustrated, should be made easily available in all cases of possible ambiguity, together with listed differences from any 'look-alike species' likely to be encountered. Ideally, the focal species should be clearly and unmistakably recognisable by non-specialist observers, so that their observations are valid contributions, and do not lead to further confusion. For most species, clear recognition may be restricted to the adult stage, but wherever possible, information on early stages and biology should also be disseminated. A readily available 'reference source', ideally a local person who is able (and willing) to confirm identifications from voucher specimens or photographs (where possible), is a major advantage.

Highly characteristic and charismatic species may especially appeal to the wider community, either as a 'local emblem' or more widely. The Eltham copper butterfly (above) was adopted as a local conservation flagship when rediscovered in outer north-east Melbourne—and the patrimony of the given common name being the suburb where it occurred fostered considerable local pride and interest amongst residents and local authority. Unusual biological features—in this case of the caterpillars feeding nocturnally, being accompanied by a specific host ant, and sampled by torchlight counts on the host plant—also increased interest and volunteer participation (New 2011). Also in Australia, conservation of the Richmond birdwing (*Ornithoptera richmondia*, Papilionidae) has depended on extensive plantings of its larval food plant, and suppression of an 'ecological trap' of an invasive vine attractive to female butterflies to oviposit, but toxic to the young larvae. Many schools participated in rearing caterpillars and monitoring vines established in their school grounds. This initiative commenced with involvement of CSIRO's young persons' 'Double Helix Science Club' in 1992, and was formalised by a ministerial launch in Queensland (1993) leading to involvement of more than 130 schools within a year. By 1996, a commercial nursery producing the food plant for distribution had sold more than 15,000 vines for planting in school yards, gardens, parklands, and regeneration areas. The butterfly has thus been encouraged to disperse and reestablish over much of its historical range as its critical resource became available—with community support throughout a long-running project (Sands and New 2013), and without which progress would have been far less. These two butterflies illustrate the contrast between (1) focusing on small isolated sites as the conservation arena for a poorly dispersing species and (2) adopting a wide landscape-level approach amenable to a large strongly flying species. The constituency of potential community aid differs accordingly—from (1) strong local support, leading to a 'Friends of the Eltham Copper' group fostering effective liaison with local authority for protection and enrichment of small isolated urban sites, to (2) a geographically wider network of concern across much of south-east Queensland and north-east New South Wales and spanning a number of local administrative areas. Both campaigns, however, depend on community initiatives, interest, and continuing support—essentially on 'citizen science', with continuing communication and recruitment of interested people. The need for public education and involvement to help mitigate insect declines has never been greater and more urgent.

8.3 Scientific Leadership

Many insects are far less 'charismatic' than butterflies, but still become important symbols for conservation. Recognition of species in many orders may not be easy, their public image non-existent, and biological knowledge restricted, so that guidance from entomologists may be necessary in detecting and monitoring their incidence and needs, and in assembling the scattered existing information and possible sources of further information and advice. 'Scientific leadership', either from a specialist in the taxon involved or with experience in practical insect conservation (or both!), can also contribute to an efficient management plan and monitoring progress, providing benefits not otherwise easily obtainable. Clearly, the best available advice is needed at all stages of an insect conservation enterprise, and that advice—whether on identification, ecology, distribution, or other theme—may be available only from an experienced worker. It should be sought at an early stage of a project—perhaps by initial query to a local museum, university department, or government agency; for many insect groups preliminary searches on 'the web' are rewarding for both information and direction.

Stoneflies (Plecoptera), especially the localised endemic species found at higher elevations, are one such group for which non-professional knowledge is rare, but have become of conservation concern in Australia. Their profile has been increased markedly by this attention. The tiny (adults up to about 12 mm long) and inconspicuous Mt. Donna Buang wingless stonefly (*Riekoperla darlingtoni*, Gripopterygidae) is known only from small streams near the summit of a single mountain in the Yarra Valley of central Victoria (p. 115). Extensive searches in similar, apparently suitable, montane habitats across the region have failed to discover it elsewhere, and it seems that the stonefly's total area of occupancy is only around 2–4 Km^2. Recent proposals to construct recreational biking trails near the summit of Mt. Donna Buang raised fears of stream damage and contamination, and advocacy for the stonefly played a central role in preventing this development of a sensitive alpine area, which is within the Yarra Ranges National Park—but clearly not protected in perpetuity. Strongly supported proposals have been made for declaration of 'critical habitat' for the stonefly. An earlier appraisal of the stonefly's conservation need (Ahern et al. 2003) noted that its range was in an area proposed for zoning for 'conservation and recreation', with risks of stream contamination and sedimentation. Larvae of the larger flightless alpine stoneflies (*Thaumatoperla* spp., Eustheniidae) are found in small streams above about 900 m, and adults of this Gondwanan genus occur streamside. *T. flaveola* occurs in the Mt. Buller/Mt. Stirling area of Victoria (Pettigrove 1991), and surveys have continued. Searches for the endangered *Thaumatoperla alpina* have innovatively trialled use of detector dogs, with considerable success as dogs alerted in the field to individual stoneflies not seen by experienced human observers (Rutter et al. 2021). The value of the dogs was amplified because, through olfactory generalisation they responded also to the closely related *Thaumatoperla flaveola*. A similar trend was found in detector dogs trained to find bumblebee nests in Britain, where a dog trained on the common *Bombus terrestris* was able to find nests of other species, including several of conservation interest (Waters et al. 2011). Earlier applications elsewhere, noted by Waters et al., include dogs detecting Gypsy moth (*Lymantria dispar*) egg masses and termite infestations.

Approaches such as the 'Action Competence for Insect Conservation' (ACIC) can potentially both increase and focus the public need and participation. Lampert et al. (2023) described this as 'an educational framework for transformational change' in contributing to insect conservation. It is founded in people's abilities to act towards solving complex problems for which decisions and actions must be taken with incomplete knowledge and based on critical thinking. An 'action' has two criteria: (1) it is deliberate and intentional and (2) it is targeted at solving a problem. The two main categories of actions are (1) direct actions, deliberate behaviour that aims to sustain insect diversity in the specified personal environment, such as garden, site, and park, and (2) indirect actions, deliberate actions that contribute to others sustaining insect diversity. Examples of each are noted in Table 8.1. Every

Table 8.1 Examples of direct and indirect actions, based on the 'Action Competence for Insect Conservation' (ACIC) framework (abbreviated from Lampert et al. 2023, and used under Creative Commons Attribution licence CC-BY-4.0)

Direct actions
Actions in personal space (garden, lawn, balcony, windowsill) that improve habitats for insect diversity at home/school/ university/workplace
Aim: Actions address the driver of urbanisation
Examples: reduce disturbance (mow lawns less frequently, leave 'wild' areas); provide diversity of flowering plants; plant pollinator-friendly plants; avoid invasive species; reduce/avoid pesticides; provide habitat and food for larval stages; use low-impact lighting
Indirect actions
Actions connected to critical customer behaviour
Aim: Actions address the driver of large-scale agriculture and forestry, and economic aspects connected to this driver
Example: buy/demand organic products and products from small-scale and seasonal production
Actions that address other people in the personal community or on a larger community scale
Aim: Actions address indirect drivers of personal perspectives; actions enable and support others to take further direct/indirect actions
Examples: Inform other people about the importance of insects, about problems of insect decline, and about actions to sustain insect diversity; ask and support others to take actions; support and engage in environmentally friendly politics, organisations, and initiatives
Actions that contribute to scientific progress
Aim: Actions address the problem of scientific gaps
Examples: Engage as a citizen scientist; support scientific organisations

action can be classified as either direct, indirect, or both, and this fundamentally simple scheme means that the ACIC scheme can be used by all relevant stakeholders. Perspectives include 'environmental' (how an action contributes to sustaining insect diversity) and 'environmental education' (with value depending on how the action contributes to developing action competence). Lampert et al. emphasised that the framework of ACIC builds on three interrelated components of action-orientated knowledge, the confidence in one's actions, and willingness to take these actions. 'Knowledge' consists of awareness of effects and causes of insect declines, the possibilities and strategies for change, and a vision for the future whilst recognising the complexity of the issues, gaps in knowledge, and relevant social perspectives. 'Confidence' reflects perception of one's personal practical skills to take actions that sustain insect diversity and that the actions are relevant, with confidence increased through learning. 'Willingness' to take action, a personal attribute, reflects positive emotions, experiences, attitudes, autonomy, and aspects such as time available and capability. Collectively, the ACIC scheme can apply to both formal educational contexts (schools and tertiary institutions at all levels) and more informal contexts such as community groups, natural history societies, and local environmental associations. It has considerable importance in enlisting and honing benefits from citizen science participation.

Exemplifying the contexts in which citizen influence can lead to significant policy amendments, the European Citizens' initiative 'Save bees and farmers' was presented successfully to the European Commission in late 2022, after collecting more than a million statements of support from European Union citizens. This decisive call for action was noted in the call for 'A new deal for pollinators' (European Commission 2023).

8.4 Citizen Science

The term 'citizen science' applies widely to the participation of non-scientist volunteers or other helpers—essentially as a corpus of support for the very limited scientific leadership that is available—in conservation survey, monitoring, and other information gathering, as well as in practical conservation management and that, with adequate supervision and continuing tuition, can furnish accurate and high-quality information. That support has been noted frequently in most 'cases' mentioned earlier, and much of the progress has depended on help from concerned volunteers able to contribute time and effort for data gathering and interpretation, under scientist guidance, and who can thus participate meaningfully. As Legge et al. (2023) commented, efforts are increasing to organise citizen science contributions into robust sampling designs, in which a variety of sampling tools can be used. The increasing use of social media for rapid communication and recording is now an integral component of citizen science activities.

Efforts of citizen scientists in Australia are important in increasing public awareness and appreciation of the values of biodiversity. The Australian Citizen Science

Association's (ACSA) definition of citizen science is 'citizen science involves public participation and collaboration in scientific research with the aim to increase scientific knowledge'. ACSA has enunciated ten principles for citizen science activities as a guide to expectations (Table 8.2). In practice, this role may be taken up by community groups or individuals and participants range from school children to retirees. Engaging school children can, for instance, take advantage of readily accessible urban habitats such as school grounds for projects such as pollinator surveys (Saunders et al. 2018). The scheme 'Pollinators in the Playground' used artificial flowers and allowed students to examine the kinds of insects that visited them (Prendergast et al. 2021). Monitoring incidence of butterflies and rearing them, as exemplified by the pivotal role of schools in conserving the Richmond birdwing, *Ornithoptera richmondia*, in Australia (p. 147) (Sands and New 2013) can also involve coordinated networks of schools.

Another facet of entomological citizen science, the programme 'Insect Investigators' engages about 50 schools in Queensland, South Australia, and Western Australia in the science of insect taxonomy and in which insect catches are examined by specialists and notable species discussed; eight schools (at August 2023) have been reported to discover previously undescribed insect species amongst their trap catches when these were examined by entomologists, one of the most recent being Year Eight students at Beerwah State High School, Queensland, finding a new parasitoid wasp (Braconidae, Miracinae, representing a group not formally recorded from Australia until 2022!) (Slater-Baker et al. 2022). Several of the above examples were discussed also by New (2022), but remain pertinent for both what they have achieved or continue to contribute and as illustration of how young people can be encouraged to notice and study insects in their local environments.

Participation by citizen groups in Australian insect conservation seems destined

Table 8.2 The ten working principles of citizen science developed by the Australian Citizen Science Association, derived from Robinson et al. (2018)

1.	Citizen science projects actively involve citizens in scientific endeavour that generates new knowledge and understanding
2.	Citizen science projects have a genuine scientific outcome
3.	Citizen science provides benefits to both science and society
4.	Citizen scientists may participate in various stages of the scientific process
5.	Citizen scientists receive feedback from the project
6.	Citizen science, as with all forms of scientific inquiry, has limitations and biases that should be considered and controlled for
7.	Where possible and suitable, project data and meta-data from citizen science projects are made publicly available and results published in an open access format
8.	Citizen scientists are suitably acknowledged by projects
9.	Citizen science programmes offer a range of benefits and outcomes which should be acknowledged and considered in project evaluation
10.	The leaders of citizen science projects take into consideration legal and ethical considerations of the project

to increase, in parallel with increased appreciation of need and the great variety of taxa and environments in which such aid can become invaluable. The novel viewpoints and activities that emerge from this broadened participation can add enormously to awareness of insect needs and how these may be satisfied.

References

Ahern LD, Tsyrlin E, Myers R (2003) Mount Donna Buang wingless stonefly *Riekoperla darlingtoni*, Action Statement no 125. Department of Sustainability and Environment, Victoria

Bennett NJ, Roth R, Klain SC, Chan KMA, Clark DA (et al.) (2016) Mainstreaming the social sciences in conservation. Conserv Biol 31: 56–66. https://doi.org/10.1111/cobi.12788

Bennett NJ, Roth R, Klain SC, Chan K, Christie P (et al.) (2017) Conservation social science: understanding and integrating human dimensions to improve conservation. Biol Conserv 205: 93–108. https://doi.org/10.1016/j.biocon.2016.10.006

Cheesman OD, Key RS (2007) The extinction of experience: a threat to insect conservation? In: Stewart AJA, New TR, Lewis OT (eds) Insect conservation biology. CAB International, Wallingford, pp 322–348

European Commission (2023) Communication for the commission to the European Parliament, the council, the European economic and social committee and the committee of the regions. Revision of the EU Pollinators Initiative. A new deal for pollinators. COM/2023/35 final

Forbes SL, Cullen R, Grout R (2013) Adoption of environmental innovations: analysis from the Waipara wine industry. Wine Econ Policy 2:11–18

Gangwani K, Landin J (2018) The decline of insect representation in biology textbooks over time. Am Entomol 64(4):252–257. https://doi.org/10.1093/ae/tmy054

Kietska G, Pryke S, Gaigher R, Samways MJ (2021) 32 years of essential management to retain value of an urban dragonfly awareness pond. Urban Ecosyst 24:1295–1304. https://doi.org/10.1007/s11252-021-1115-5

Lampert P, Goulson D, Olsson D, Piccolo J, Gericke N (2023) Sustaining insect biodiversity through action competence—an educational framework for transformational change. Biol Conserv 283:110094. https://doi.org/10.1016/j.biocon.2023.110094

Legge S, Rumpft L, Garnett ST, Woinarski JCZ (2023) Loss of terrestrial biodiversity in Australia: magnitude, causation, and response. Science 381:622–631

Matthews RW (2000) Developing life science instructional materials using a parasitic wasp, *Melittobia digitata* Dahms (Hymenoptera: Eulophidae): a case study. In: Austin AD, Dowton M (eds) Hymenoptera. Evolution, biodiversity and biological control. CSIRO Publishing, Collingwood, pp 427–433

Matthews RW, Flage LR, Matthews JR (1997) Insects as teaching tools in primary and secondary education. Annu Rev Entomol 42:269–289

Merritt DJ (2023) Entomological tertiary education in Queensland and Australia. Aust Entomol 50:179–186

Moore NW (1997) Dragonflies: status survey and conservation action plan. IUCN/SSC Odonata Specialist Group, Gland and Cambridge

New TR (2011) Butterfly conservation in South-Eastern Australia: progress and prospects. Springer, Dordrecht

New TR (ed) (2012) Insect conservation: past, present and prospects. Springer, Dordrecht

New TR (2015) Insect conservation and urban environments. Springer, Dordrecht

New TR (2022) Insect diversity, declines and conservation in Australia. Springer Nature, Cham

Pettigrove V (1991) An investigation of the distribution and ecological requirements of the stonefly *Thaumatoperla flaveola*, Rural Water Commission of Victoria, Report no. 104

Prendergast K, Vanderstock A, Neilly H, Ross C, Pirotta V, Tegart P (2021) Potential and pitfalls of citizen science with children: reflections on pollinators in the playground project. Austral Ecol 47(2):189–195. https://doi.org/10.1111/aec.13031

Robinson LD, Cawthray JL, West SE, Bonn A, Ansine J (2018) Ten principles of citizen science. In: Hecker S, Haklay M, Bowser A, Makuch Z, Vogel J, Bonn A (eds) Citizen science: innovation in open science, society and policy. UCL Press, London, pp 27–40. https://doi.org/10.14324/111.9781787323339

Rutter NJ, Mynott JH, Howell TJ, Stukas AA, Pascoe PC (et al.) (2021) Buzzing with possibilities: training and olfactory generalization in conservation detector dogs for an endangered stonefly species. Aquat Conserv 31: 984–989. https://doi.org/10.1002/aqc.3531

Sands DPA, New TR (2013) Conservation of the Richmond birdwing butterfly in Australia. Springer, Dordrecht

Saunders ME, Roger E, Geary WL, Floret M, Welbourne DL (et al.) (2018) Citizen science in schools: engaging students in research on urban habitat for pollinators. Austral Ecol 43: 635–642

Sitar G-M, Rusu AS (2023) The impact of environmental educational programs in promoting insect conservation awareness: a scoping review. J Educ Sci 47:1. https://doi.org/10.35923/JES.2023.1.05

Slater-Baker M-R, Austin AD, Whitfield JB, Fagan-Jeffries EP (2022) First record of miracine parasitoid wasps (Hymenoptera: Braconidae) from Australia: molecular phylogenetics and morphology reveal multiple new species. Austral Entomol 61:49–67. https://doi.org/10.1111/aen.12582

Suh AN, Samways MJ (2001) Development of a dragonfly awareness trail in an African botanical garden. Biol Conserv 100:345–353. https://doi.org/10.1016/S0006-3207(01)00038-6

Vanderstock A, Grandi-Nagashiro C, Kudo G, Latty T, Nakamura S (et al.) (2022) For the love of insects; gardening grows positive emotions (biophilia) towards invertebrates. J Insect Conserv 26:751–762. https://doi.org/10.1007/s10841-022-00419-x

Waters J, O'Connor S, Park KJ, Goulson D (2011) Testing a detection dog to locate bumblebee colonies and estimate nest density. Apidologie 42:200–205. https://doi.org/10.1051/apido/2010056

Chapter 9
Towards an Agenda for Insect Conservation in Australia

9.1 Introduction: Current Conservation Perspective and Need

Continuing investigation and documentation of Australia's insect fauna is leading to a fuller picture of the vast diversity present and its high level of endemism and evolutionary interest—and also to the needs for its conservation to be accelerated and focused to prevent many losses from occurring over the coming decades. Compared to the longer history of natural history interests in Britain, western Europe, and North America, interest in insect conservation in Australia is relatively recent—and concern for insects even more so in the face of concerns for iconic marsupials and other vertebrates, and realisation that Australia's 'extinction record' for these most popular animals is as an unenviable world leader. Writing more broadly on Australia's biodiversity losses, Legge et al. (2023) recognised that 'the situation is dire' but suggested that Australia has a number of distinct advantages that could help stem losses and enable recovery (Table 9.1), in conjunction with public goodwill.

Much of the earlier history of insect conservation in Australia, summarised by New and Yen (2012), reflects that there have never been large numbers of lepidopterists, coleopterists, or other hobbyists and insect collectors in Australia, so that faunal knowledge has not paralleled that accumulated from these leisure interests elsewhere—but, notwithstanding this, the relatively few dedicated collectors provided sound foundation of material in major museums. The great majority of butterflies in museums, for example, are the depositions of non-professional collectors (Moulds 1999), including specimens from the nineteenth century on. However, many insect orders have rarely or never attracted attention from hobbyists. For these taxa, accumulated data are sparse, and largely the outcome of attention from a few specialist devotees.

T. R. New, *Insect Conservation in Australia: Why and How*,
https://doi.org/10.1007/978-3-031-66631-5_9

Table 9.1 The 'distinct advantages' of Australia in helping to stem biodiversity losses and enable recovery, suggested by Legge et al. (2023)

1	The continent of Australia is a single nation, making legislation, policy, and management responses theoretically simple
2	It is politically stable and wealthy, which should allow consistent, strategic, and long-term responses to conservation problems
3	It has a low population density, which should cap pressure on the Environment
4	Many Indigenous peoples retain a culture of caring for Country, and continue to manage large areas of the continent

Development of interest in insect conservation in Australia reflects a variety of origins and uneven concerns from different interest groups, initiated variously by field naturalists, professional entomologists, and entomological societies (perhaps the leading impetus, flowing from experience and knowledge), individual academics (mostly for particular cases, some then broadening interest), establishment of conservation organisations (historically with rather minor roles, but recently increasing), and government agencies (again with little historical interest, but more engaged since the 1980s) (New and Yen 2012).

The most significant advances in understanding have come from a relatively few high-profile 'flagship species', and mark notable progress since Key (1978) emphasised the close relationships of insects to their habitats and suggested their values in monitoring habitat condition. Establishing a theme that continues to the present (Taylor et al. 2018), Key urged the need for a programme that will 'embrace the reservation of samples of all the ecosystems' whilst admitting that this aim is 'slightly utopian'. That aim has been recapitulated repeatedly, on the grounds that it is 'the only sound approach to conservation of a very large and diverse fauna' (Hill and Michaelis 1988). Other than the Lord Howe Island stick insect (p. 86) and the wider Lord Howe fauna, Key noted only representatives of Odonata and Orthoptera as needing conservation, a taxonomic perspective broadened by respondents to a national survey by Hill and Michaelis (1988), which collectively noted 260 species of invertebrates (28 of them not insects) that merit attention to determine their conservation needs. That breadth of concern was enlarged in the national overview by Yen and Butcher (1997), so that representatives of many insect orders emerged as needing further evaluation or conservation attention. More recent guidance has come from concerns expressed by state-based entomological societies and the national Australian Entomological Society (AES), whose conservation committees or representatives have raised many important issues, some involving listed species, nominating additional species for listing, flagging environmental changes, and contributing to community engagements and the activities of other interest groups in conservation programmes. The eight states and territories of Australia each have independent legislations and priorities for conservation, with national concerns embraced progressively under the Commonwealth's Environment Protection and Biodiversity Conservation Act (EPBC Act). Thus, the AES Conservation Committee has members representing all eight regions.

The priority importance of insect conservation was noted for New Zealand in an appraisal of the critical issues facing entomology in that country (Lester et al. 2014). Those issues (Table 9.2) included 'conserving indigenous invertebrate diversity', and several themes directly addressing key threats, most notably from the impacts of invasive species. No such parallel set of priorities has yet been elected for Australia, but the listing in Table 9.2 is likely to overlap considerably with Australian needs. One major difference in emphasis reflects New Zealand's historical absence of mammals (other than bats), and the major threat to native insects from introduced mammal predators, with their suppression a key focus of much insect conservation.

9.2 Priorities and Progress

Suggestions and nominations of threatened insect species since then have established that both concerns and needs for insect conservation are considerable, and increasing as awareness of the extent of Australia's insect fauna and its vulnerabilities also grows. Rather few species are 'listed' as threatened under the EPBC Act (Table 9.3), and historical disparities in appraisal of the same species under different legislations have weakened some cases through lack of consensus over conservation need. The species listed are likely to represent only a tiny proportion of those that merit equivalent notice. Most of those species have been discussed extensively, and some have a strong public profile and have become symbols for wider conservation needs and advocacy; approved conservation advice published for each listed species sets out the case for conservation, the grounds for listing, and the major conservation needs. As examples from this list and other species signalled under more local legislations: (1) the Golden sun-moth (*Synemon plana*, Castniidae) is a flagship species for threatened native grasslands on the south-east mainland, and is listed under EPBC and legislations for all range areas—Victoria, New South Wales, and the Australian Capital Territory; (2) the Eltham copper butterfly (*Paralucia*

Table 9.2 The nine topics selected as 'critical issues' facing New Zealand entomology by Lester et al. (2014), listed in no ranking priority

1.	Conserving indigenous invertebrate diversity
2.	Limiting predator pressure exerted by exotic vertebrate predators
3.	Limiting the effects of invasive invertebrates, especially *Vespula* wasps in honeydew beech forests
4.	Understanding the influence of pesticides and pathogens on honeybees
5.	Maintaining and enhancing efficient biosecurity systems to keep out invasive pests and diseases
6.	Enhancing support for taxonomy
7.	Enhancing entomological teaching, training, and support for entomology graduates
8.	Utilising the hidden knowledge of retired entomologists
9.	Supporting amateurs in entomology

Table 9.3 Insect species listed formally for conservation interest as 'threatened' under Australia's Commonwealth Environment Protection and Biodiversity Conservation Act 1999, at October 2023 (Conservation advice for each species, setting out the background, case, and needs, is included on the EPBC web site: https://www.dcceew.gov.au>environment>epbc)

Status	Species and effective date
Critically endangered	*Dryococelus australis*, Lord Howe Island Stick insect, July 2002
	Micropathus kiernani, Francistown cave cricket, Southern sandstone cave cricket, January 2014
	Pseudococcus markharveyi, *Banksia montana* mealybug, May 2018
	Cormodes darwini, a clerid beetle, September 2023
	Hoplogonus bornemisszai, Bornemissza's stag beetle, January 2009
	Promethis sterrha, a tenebrionid beetle, September 2023
	Argynnis hyperbius inconstans, Australian fritillary butterfly, August 2017
	Hypochrysops piceatus, Bulloak jewel butterfly, September 2023
	Ogyris subterrestris petrina, Arid bronze azure butterfly, August 2015
	Hesperocolletes douglasi, Rottnest bee, July 2019
	Leioproctus douglasiellus, a short-tongued bee, May 2013
	Neopasiphae simplicior, a native bee, January 2009
Endangered	*Austrocordulia leonardi*, Sydney hawk dragonfly, April 2022
	Thaumatoperla alpina, Alpine stonefly, March 2011
	Keyacris scurra, Key's matchstick grasshopper, October 2022
	Trioza barrettae, *Banksia brownii* plant louse, July 2019
	Lissotes latidens, Broad-toothed stag beetle, December 2002
	Antipodia chaostola leucophaea, Tasmanian chaostola skipper, August 2010
	Oreixenica ptunarra, Ptunarra brown butterfly, January 2014
	Paralucia pyrodiscus lucida, Eltham copper butterfly, May 2016
	Phyllodes imperialis smithersi, Pink underwing moth, December 2002
	Trisyntopa scatophaga, Antbed parrot moth, December 2016
Vulnerable	*Hoplogonus simsoni*, Simson's stag beetle, March 2011
	Hoplogonus vanderschoori, Vanderschoors stag beetle, March 2011
	Oreisplanus munionga larana, Marrawah skipper, March 2011
	Paralucia spinifera, Bathurst copper, July 2000
	Synemon plana, Golden sun-moth, July 2000

pyrodiscus lucida, Lycaenidae) (EPBC and its home state, Victoria) awakened interest in the vulnerabilities of localised insects to urban development and the critical importance of tiny occupied sites; (3) the Richmond birdwing butterfly (*Ornithoptera richmondia*, Papilionidae) (Queensland) demonstrated the benefits from community participation in a range-wide habitat rehabilitation exercise (Sands and New 2013); (4) the Ancient greenling damselfly (*Hemiphlebia mirabilis*, Hemiphlebiidae), a putative 'living fossil', was the first aquatic insect to gain conservation attention, including successful searches for additional populations; and (5) the Lord Howe Island stick insect (*Dryococelus australis*, Phasmatidae) demonstrated how captive breeding (here of a large and 'showy' species) from a very small source stock may be instrumental in 'saving a species', with potential for later release to the wild once major threats have been mitigated. Lessons learned from these taxa have helped to guide wider interests and conservation progress, as noted below.

Any such listing is 'dynamic' and can change as knowledge accumulates—hence the revised status of the Golden sun-moth (*Synemon plana*), formerly assessed as 'critically endangered' but now 'vulnerable'. The Rottnest bee (*Hesperocolletes douglasi*) was gazetted in Western Australia as 'presumed extinct' in 1994, but a single female was collected in a *Banksia* woodland remnant near Perth in 2015. Further searches failed to reveal any further specimens, but a small population of the bee may persist (Arnold et al. 2019).

Several other species on the EPBC list also have very small known distributions. The cave cricket *Micropathus kiernani* is known from a range area of less than 0.1 Km^2, for example. Two beetles from the Lord Howe Island area, a designated World Heritage Area, are known from single localities of less than 1 ha (*Cormodes darwini,* Cleridae) and 4 Km^2 (*Promethis sterrha*, Tenebrionidae) and were amongst the search targets of well-organised citizen science surveys on their native Blackburn Island (Reid and Hutton 2019; Reid et al. 2020), but without discovery of further populations. Their loss may be due to rodent predation, weed invasions, or other alien species impacts. Several other listed insect species are tied closely to particular food plants—so that the major concern for the plant louse *Trioza barrettae* is the decline of its specific host—a circumstance that may have already led to co-extinction of *Pseudococcus markharveyi* with the decline of *Banksia montana* in Western Australia (Moir 2021). For both these herbivores, synergistic impacts of *Phytophthora* fungus dieback of the host plants and severe fire are key drivers of loss.

Attempts to assess the extent of conservation needs of larger taxonomic groups depend on seeking consensus and advice from the greatest number possible of informed and interested people, and this has so far proved possible only for butterflies. A 'National Action Plan' for Australia's butterflies (Sands and New 2002) was founded in a series of workshops across Australia, attended by most interested hobbyists and lepidopterists, agency personnel, and others, considering and debating the conservation status of all taxa, and—where believed necessary—possible remedial measures. A more recent workshop suggested the most threatened butterfly species (Geyle et al. 2021) and evaluated the risk of extinction by 2040 of 26 priority species listed by consensus and the likely causes of this. Most are narrow-range endemics, limited to a single state or territory, but only one (the Laced fritillary (or Australian fritillary, *Argynnis hyperbius inconstans*, Nymphalidae) was assessed as particularly severely endangered, and a further four were estimated to have a moderate to high risk of loss.

Most other Australian insects have not undergone any such comprehensive appraisal. However, Hawking's (1999) categorisation of Odonata—as another reasonably well-known order—revealed three concerns that are likely to be far more widespread. These were (1) the large number of species (76, or 24% of the fauna) believed to be of conservation concern by fulfilling the IUCN criteria: 37 species were 'endangered'; (2) a large number (39) of 'priority species', categorised as 'near threatened' or of special biological interest; and (3) those species (84) insufficiently known for evaluation and for which Hawking urged immediate investigation, supposing that some may be near extinction, or even already extinct. Even for these most popular insect groups, considerable uncertainties persist; for other taxa, conservation need can be only conjecture as a measure of environmental changes and knowledge of—at most—a few included species.

Despite the welcome publicity and expressions of concern arising from the few 'listed' species', wide consensus exists that they cannot realistically represent the vulnerabilities of much of the remaining insect fauna—and also that the prospect of redressing this need in sufficient detail is utopian. Even providing focused recovery plans for the few listed species has proved difficult, as has revising these for taxa for which any plans do exist, and pursuing the plans to practice. Calls for further species listings to broaden the current scope to seek coverage of Australia's major bioregions (Taylor et al. 2018, p. 168) reflect the needs for listed species to have greater 'umbrella' coverage, drawing attention to the great variety of ecosystems and threats to them, and that each may support taxa (and assemblages and communities) that include narrow-range endemic insect taxa worthy of conservation on taxonomic and ecological grounds. In parallel, Australia's extensive protected areas throughout the continent provide much opportunity for badly needed surveys to determine and enumerate their insect fauna, to detect taxa of particular conservation interest, and to assess their vulnerability in the face of environmental disturbances and climate change. In some places, cultural sensitivities may arise over access to particular sites and working in them, and liaison with First Nations people may be needed to obviate these. 'Land tenure' defines legal land uses over two major categories: (1) freehold land is owned by the landholder in perpetuity, and (2) Crown land includes numerous tenure categories owned by the state (the 'Crown') and leased or licensed for particular land uses. These were estimated to cover 29% of the continent (freehold) and 38% (leasehold), with protected areas covering 20% (Kearney et al. 2021), but most threatened species (over all taxa) occur across multiple tenures, necessitating coordinated management across different tenure systems and levels of security.

Notwithstanding any protected area status, it is clear that some unique Australian ecosystems are under anthropogenic or other threats, and that opportunity to document their biota, examine their likely fate, and assess whether any conservation moves may be possible may disappear within—at most—a few decades. The restricted alpine areas in the south-east, for example, are faced with contraction as climate change proceeds, but this is exacerbated by current expansions of recreational activity, such as winter sports and increased summer tourism and their associated environmental impacts in providing access and facilities. Locally endemic insects are threatened by direct habitat loss, the impacts of invasive species driven upwards by climate warming, possible changed synchrony with critical resources, and increased human pressures throughout the year. As another example, localised coastal ecosystems—such as mangrove swamps and dunes—are vulnerable to rising sea levels and increased human recreational needs. Again, urban expansion continues to envelop lowland native grassland areas, for several decades considered as Australia's most endangered ecosystems (Kirkpatrick et al. 1995). Diverse native forests are still vulnerable to timber demands—and the forestry industry of the south-east indeed furnishes livelihoods for many people, illustrating the dilemma of harmonising human wellbeing with acute conservation need. Replacement plantation establishment based on native tree species, whether monoculture or not, may help salve human economy but is far less likely to re-create the heterogeneity needed by diverse invertebrate communities and much other Australian wildlife. In short, many of Australia's key insect habitats are under some sort of threat, with the outcomes highly uncertain, but awareness that many native insects are narrowly distributed and ecologically specialised suggesting that losses of species may be both numerous and undetected. A recent sobering evaluation of vegetation loss found that more than 7.7 million hectares of potential forest and woodland habitat in Australia were cleared over 2000–2017, since the EPBC Act came into effect in 2000 (Ward et al. 2019); more than 93% of this had not been referred for assessment, so did not undergo the scrutiny required under the Act. The inevitable inference was that the Act is ineffective in protecting the habitat and ecological communities of numerous terrestrial animals—including unknown numbers of insects. Summaries of insect conservation needs and possible priorities in some key Australian habitat types (grasslands, forests, inland waters) exemplify the diversity of taxa and vulnerabilities that confront and challenge conservation managers (New 2018, 2019, 2020), and the practical lessons that may be learned from elsewhere.

In formal conservation status terms, most Australian insects are currently unassessed against the Red List criteria and in principle are 'data deficient'. Many are known from few specimens (some from single individuals) and localities, and some have not been seen for many years since they were described. Targeted searches for many have not led to rediscovery. A possibly extreme case is for a small moth *Lophocorona melanora* (Lophocoronidae) known from a single individual captured in the Black Mountain Reserve, Australian Capital Territory, in 1949. The reserve is perhaps the most heavily surveyed area for insects in the country, as being near the headquarters of the Australian National Insect Collection (ANIC), but many subsequent efforts have failed to find *Lophocorona* again. For this, and numerous other species (most of which have not been sought in equivalent detail, as being known from more distant or less accessible sites), it is uncertain whether it still exists—if so, and again in parallel with others, it is highly elusive and purportedly very rare.

The incidence of numerous 'rare' insects in Australia complicates defining conservation priorities and allocating the limited resources available. Much knowledge of insects over large areas of Australia away from the most heavily populated eastern coastal fringe and the south-west corner has come from sporadic expeditions rather than long-term continuing surveys. The term 'expeditions' is deliberate—distance, remoteness, seasonal restrictions from weather, indigenous cultures, funding, and travel all add dimensions of insect surveys far different from many undertaken in more populated areas and, with a few exceptions, have meant that knowledge is based on short-term 'snapshot' inspections alone. The logistics of several well-organised such trips undertaken by CSIRO (now more specifically ANIC) staff were described by Upton (1997). The historical records from these are a vital part of Australia's entomological history, but revisiting remote sites to examine changes is itself sometimes difficult. In one of the most extensive series of surveys undertaken, K.H.L. Key and his colleagues documented grasshoppers over 223 systematic surveys from 1936 to 1989, with his notebooks recording the findings over about 2700 days of survey effort across much of the continent (Kearney et al. 2020). Kearney et al. attempted to replicate the sampling methodology (described in detail in the notebooks) used for three such expeditions in New South Wales (1946–1949) to investigate the grasshopper fauna at 45 locations in 2019. Even re-finding the precise sites inspected by Key was sometimes difficult: Key's data were based on odometer readings of distance by road from a fixed point, accompanied by written descriptions of terrain and vegetation. Richness was generally higher in the Kearney surveys, and some genera showed considerable changes in range and incidence, some linked with the previous year's rainfall. Key's 230 field notebooks are a rich source of information as baseline data for further Orthoptera surveys.

Some species widely believed to have become extinct but either targeted for rediscovery or sufficiently distinctive to be recognisable if found by chance have indeed been found. The Ancient greenling damselfly, *Hemiphlebia mirabilis* (Hemiphlebiidae), was previously known from a few small reedy lagoon sites in the Goulburn Valley and Yarra Valley areas of Victoria, where it was believed to have been exterminated by agricultural activities such as drainage for pastoralisation, leading Davies (1985) to comment that 'the cow must be *Hemiphlebia*'s worst enemy'. The tiny short-lived damselfly, the only representative of its presumed very primitive family and then regarded as a 'living fossil' with no recent extant relatives, was a high global priority for investigation. The discovery of a population in Wilsons Promontory National Park, Victoria (Davies 1985), led to study of its biology (Sant and New 1988) and, in due course, discovery of further populations in Victoria and Tasmania, including a very strong population in western Victoria (Cordero-Rivera 2014, 2016) that has allayed fears of imminent loss. With an estimated 1 million individuals in that population, the then designation of 'critically endangered' was clearly not justifiable (Cordero-Rivera 2016).

The progress with documenting the status of *Hemiphlebia* exemplifies a situation probably more widespread in insect conservation, of a species believed to be extinct or severely threatened over a putatively narrow range being given conservation priority, but with additional surveys then revealing it to be present elsewhere or to be more secure than initially supposed. There is thus commonly uncertainty over the reality of claims of conservation need. But, as for *Hemiphlebia*, much of the incentive for exploration for notable insect species flows directly from their purported scarcity and, for some, their formal listing obliging or facilitating that exploration. Without such recognition—even if later found to be unnecessary—knowledge of several such species in Australia could not have advanced to the state of making a more informed assessment of need. This is perhaps the best outcome from listing a species: to demonstrate unambiguously that it should not have been listed in the first place—but to recognise that without listing its true status would not (perhaps could not) have been established!

Such species can in due course be 'delisted', so removed from competing for scarce logistic support. Another category of candidates for delisting are species for which conservation actions have succeeded and led to security, with management obviating or minimising future risk of extinction. Some of these species may have received long-term and costly management—and there is often suspicion that threats might recur and necessitate future conservation attention. Sands and New (2002, also New and Sands 2003) suggested that species delisted as the outcome of successful conservation management should be monitored to detect any future emerging need and protect the earlier conservation investment, rather than be 'forgotten'; they suggested the term 'rehabilitated species' to signify a special status for these as having benefitted from conservation attention. The Richmond birdwing butterfly (p. 147), now again present over much of its historical range because of removal of a toxic alien food plant and increased resource availability through planting of nursery-grown stock of the preferred native larval food, was regarded as a possible candidate for this status. The Lord Howe Island stick insect (*Dryococelus australis*) was amongst the species discussed in developing criteria for assessing the 'Green Status' of species recovery (Akcakaya et al. 2018), with the major gains and high recovery potential flowing directly from captive breeding and likelihood of release across the entire historical range.

Again as for *Hemiphlebia*, assessing the status and population size of candidates for conservation priority reflects the extent of biological knowledge of the species, and that repeated surveys may be needed even to detect a highly seasonal species, or one whose activity is governed by weather or by other local conditions. 'Apparency' of many insects is often very variable. The Golden sun-moth (p. 119), as with *Hemiphlebia*, was initially believed to be extremely scarce, known from few native grassland patches on the south-east mainland, and regarded as critically endangered. Detection is by day-flying males, as females are largely inactive, and counts of moths seen indicated abundance: however, many early surveys on apparently suitable sites did not reveal the moth. More detailed investigation showed that the moths were active for only a few hours around the middle of the day, and flew only under rather restricted weather conditions (Gibson and New 2007). Surveys at other times of day or in unsuitable weather gave impressions of absence. The moth's complex biology gives other evaluation problems. At any site, the sun-moth's flight season can be up to several months from late spring onwards, but over that time moths emerge continually from subterranean pupae, with each adult living for only a few days: any single visit during the flight season can record only an unknown proportion of the total population. Further, because it is still unclear whether a moth generation takes 1, 2, or even 3 years at any site, even a highly laborious and carefully repeated series of counts in any 1 year may encounter only an unknown cohort of the larger resident population. However, increased detectability and more effective surveys for males have shown *S. plana* to be far more widely present than earlier supposed, and conservation planning accepts that population sizes are problematical to assess. Counts of pupal cases projecting from the ground have been suggested as a cumulative record of moth emergences over a season and allowing assessment of moth breeding in different habitats (Richter et al. 2012). The empty cases persist for at least 3 weeks, with 95% found after 24 days, after which detectability decreased; all had degraded after 50 days from first detection.

Most insects are less problematical to survey, but each may have biological features that aid or hamper study through behaviour, seasonal variations, and general 'detectability'. The Eltham copper butterfly, above, can be monitored from both adult and larval stages. Adults are conspicuous and active, and cannot be confused with other Lycaenidae on the sites they occupy—although sightings of a related species in Victoria have occasionally been reported as possible new findings of *Paralucia pyrodiscus lucida*: a reminder that it is necessary to characterise focal species carefully and, as far as possible, unambiguously in order to validate survey results, and from time to time investigate possibly misleading reports. The adult butterflies can be assessed by standard point or transect counts on the small occupied sites. Larvae are nocturnal, and are escorted up the *Bursaria* food plants by mutualistic ants, moving from the ant nests at the plant base in which they shelter by day. The glossy ants and the butterfly larvae are visible by torchlight, and it is feasible to determine which individual plants are used, and the number of caterpillars associated with each, helping to map their distribution on sites. It is relatively unusual to be able to independently and reliably survey two generational stages of a threatened insect species. For many other butterflies, only the adult stage is amenable to counts to assess intergenerational change in abundance.

The lessons learned from the few 'flagship' species of insects in Australia, and the consequent improvements in conservation management for each, help to define a pattern for future insect conservation in Australia. Whether in the field or under captive breeding conditions, the effort required is substantial, and the wider formal demands obliged by 'listing' may by far exceed the capability available. Pragmatically this situation seems unlikely to change, and a broader perspective is needed to both streamline the rather cumbersome 'listing process' and select the optimal species to be recognised in this way.

9.3 More Species Listings

However, advocacy continues to promote listing of more insect species in order to gain greater representation of the fauna. Both the wisdom and practicality of this approach also continue to be debated, with reluctance based on incomplete knowledge—so that many potential listings of insect species are accompanied by doubts over their validity and real priority. The many benefits of highlighting the plight of individual species by formally listing them for recognition flow from their tangibility—they are entities to which managers, the public, funding bodies, and others can relate easily, much as referring to species such as the Helmeted honeyeater, Koala, or Murray cod, each an unambiguous taxon and significant conservation focus with popular support for its wellbeing. Less easy to grasp is the wealth of other insect life that may accompany a focal insect species in the same habitat, and that the species itself may not be easy to recognise, be small and inconspicuous, its identity difficult to confirm by a non-specialist, and its biology and distribution largely unknown. One respondent to the survey by Hill and Michaelis (1988) commented that most species in Australian museum collections are represented by specimens from fewer than five localities, and many entomologists were unwilling to nominate such taxa for attention because of uncertainties over their status and needs. In addition, it is unlikely that Australian published literature or more informal reports will directly address the species in any detail. Simply confirming the identity of the taxon and establishing its basic life history and main requirements can become an unexpectedly difficult and protracted task. Any available 'expert advice' should be sought early in this discovery process. Each listed insect is an ambassador for many others, perhaps equally deserving of attention but not yet recognised. However, debate continues over whether the apparently large numbers of deserving insect species should all be listed in this way, with a series of advantages and disadvantages summarised by Braby (2018, Table 9.4), or whether a carefully selected suite of taxa for listing might be a more rewarding approach and better use of limited logistic support. In some cases, lack of listing may thwart conservation: the Enigma moth (p. 190) has been assessed as 'Critically Endangered' under IUCN guidelines but is not yet (October 2023) listed under Australian laws and so has no legislative protection that

Table 9.4 Advantages and disadvantages of nationally listing invertebrates for species-focused conservation in Australia, extracted and adapted from text by Braby (2018)

Advantages
1. Explicit recognition that the species may be near extinction and requires protection and responsive management
2. Provides a snapshot of the variety and numbers of taxa of conservation concern
3. Provides justification for funding of future research to clarify biology and evaluate conservation status and needs
4. Identifies key threatening processes which may extend more broadly to the community or biotope of the species
5. Raises public profile of the species, which may then become a 'flagship' of broader conservation value
6. Places onus on developers to ensure that their activities will not have adverse impacts on the species
Disadvantages
1. Listing process is viewed as an end point rather than the initial facilitator of conservation action
2. The delisting process is difficult or not possible, despite evidence that listing is not needed
3. Lists are inaccurate, and credibility may be lost from incomplete, poor, or misleading information
4. Lists are biased so that listings do not represent overall biodiversity at risk of extinction
5. Wide discrepancy in numbers of species listed amongst different jurisdictions, reducing credibility
6. Lack of alignment between legislation in different jurisdictions or between Commonwealth—State/Territory legislations, again reducing credibility
7. Non-professional entomologists and citizen scientists are discouraged from contributing data on listed species, due to prohibition of collecting specimens

would support continuing surveys and monitoring, and conservation measures and management of its restricted coastal dune habitat on Kangaroo Island.

As elsewhere in the world, responsible and satisfactory outcomes from study of individual listed insect species in Australia are elusive, and any 'quick fixes' are unusual. Much management is to some extent 'experimental' because, whilst general principles are well-established, the idiosyncrasies of each species can complicate progress. The most beneficial measure has been to assure security of tenure and long-term habitat protection and, where needed, remediation for significant occupied sites or habitat patches. Without site security as an initial basis, pursuing conservation (of any taxon, not just insects) may be futile. In extreme cases in which a key site is scheduled for imminent development, it may be necessary to consider some form of 'rescue' or 'salvage' translocation by removing the insects before they are destroyed (p. 124) and releasing them elsewhere. This presupposes that a suitable and secure receptor site is available, and that the methods used are understood. However, change of land tenure to accommodate such needs can be expensive, and perhaps especially difficult to achieve for a poorly understood and publicly obscure insect, and for which appeal for funding may not be competitive.

The dilemma with additional listings is that (1) on the one hand they demonstrate the massive diversity and needs of Australia's insect species (and long lists may be impressive to 'politicians'!) but (2) on the other hand they reveal the massive and enduring shortfall of capability and funds to attend even the most needy species. Recent insect species nominations for conservation listings in Australia have reflected an ad hoc process involving individual species rather than a more comprehensive or reasoned selection from deserving candidates and have resulted largely from the zeal and concern of individual people or community groups. Many entomologists could suggest species amongst 'their' groups that may be equally deserving candidates. The few listed species are thus a biased subset of need, led by Lepidoptera, with needs for butterflies by far the most conscientiously evaluated. However, contributors to the survey of likely threatened insect species undertaken by Hill and Michaelis (1988) in addition suggested species of Thysanura, Ephemeroptera, Odonata, Plecoptera, Orthoptera, Phasmatodea, Hemiptera, Coleoptera, Mecoptera, Diptera, and Hymenoptera, collectively a far wider consideration of concern. Further bias occurred in that listing—56 species of Drosophilidae (Diptera) were listed, almost all poorly known taxa mostly from Queensland rainforests, but an expression of the breadth of possible vulnerability considered likely by a specialist in these flies. The species shared the same wet forest habitats, some depending on limited 'microhabitats' such as decaying flowers, and threats from clearing and degradation, but no specific ecological information was available for most of those flies. There is little reason to doubt that multiple species in many other diverse insect families would prove equally vulnerable, if they were ever appraised.

A more balanced expression of need resulted in an 'Advisory list' of invertebrate species that were considered threatened, poorly known, near threatened, or extinct in Victoria (DSE 2009) as species that might be considered in planning applications or for future conservation priority: of 102 insect species listed, 35 had been given more formal protection under the state's Flora and Fauna Guarantee Act 1988. A longer compilation of insects of conservation interest (Yen and Butcher 1997) included many species that are still very poorly known and brought to attention because of their apparent rarity. In essence, though, despite their immense value in signalling need and concern, these longer lists also convey the helplessness of constructive progress at this level and difficulties in gaining consensus on the real priority species to attend.

Historically, a widespread highest priority for attention amongst listed species has been to attempt to save those that are of greatest risk of extinction, and so for which conservation has been deemed most urgent even though chances of success appear small and for which investment may be large. However, it has also been suggested that such cases should not be heeded, because those chances of success are minimal and the resources needed should not be 'wasted'.

An alternative approach is to concentrate on species that still have a more realistic prospect of salvation, and for which conservation effort is more likely to succeed, with the important side benefit of displaying success to funding agencies and other possible sympathisers. More subjective priorities may also occur, perhaps more controversial but also more practicable and in part reflecting difficulties of setting scientifically convincing priorities from the little evidence available. As examples: (1) previous successful experience may encourage selection of related species for attention—the knowledge base for butterflies, for example, may predispose towards selecting further butterfly species rather than opting for an entirely different and unfamiliar taxon, which may not be as competitive in public opinion for support (the counter is that it might be better in terms of conserving evolutionary diversity to ignore butterflies in favour of other, less-heeded insects) and to consider features such as phylogenetic isolation If a threatened species is taxonomically isolated, such as being the only member of its family, many biologists argue that its loss has far greater significance than 'just another butterfly' from a diverse family with many secure taxa, and this reasoning was largely responsible for giving the Ancient greenling damselfly (p. 158) global priority status amongst Odonata); (2) a 'flagship' species may draw attention to the plight of its close relatives, perhaps with similar needs and biology. Interest in the Golden sun-moth (*Synemon plana*, p. 146) was in part responsible for wider interest in the status and conservation needs of other endemic and localised Castniidae on grasslands; several such species have high conservation priority in Western Australia or south-east Australia; and (3) local enthusiasm and knowledge or experience with a particular taxonomic group may be instrumental in garnering political and public support for deserving members of that taxon.

Objectively, any system of setting priorities amongst threatened species may lead to debate and strongly expressed opinions—many in the knowledge that the logistic support allocated to any single species is likely to deprive other equally deserving taxa of attention and that the 'best' choices should be made. In essence, any such system involves triage, whereby different criteria may elevate some taxa over others. But, as inferred above, criteria may extend beyond the species themselves to consider the cost of the management needed and the likelihood of success (New

2009, Braby 2018), essentially to maximise returns from the conservation investment.

A somewhat different focus on selecting priority species emphasises ecological representation to a greater extent, employing notable species as flagships for regions or unusual habitats of wider conservation concern and significance. Taylor et al. (2018) pointed out that many of Australia's major bioregions have no formal 'insect representatives', in part reflecting that most entomological interests are centred in temperate coastal Australia. In particular, very few species have been listed from the remote arid areas, and 36 IBRA regions have no insects listed from them. To improve this imbalance, Taylor et al. suggested a more systematic approach for each of the 89 IBRA (Interim Biogeographic Regionalisation for Australia) regions of selecting a relatively small set of 'flagship taxa'—comprising threatened and/or iconic species of high scientific and/or social or cultural value which are given priority (and perhaps listed) for attention. The categories are not mutually exclusive—a flagship species may be both threatened and iconic and have both scientific and social significance. The wider ramifications of this approach could lead to improved management of many key habitats and threatened ecological communities, as well as addressing key threatening processes. The strategic approach implied is shown in Fig. 9.1, with community engagement a central pillar of progress.

Threat management and prevention is central to insect conservation at any level and in any habitat. Some measures undertaken primarily for other conservation purposes may have considerable values for insects. The provision of predator-free areas for Australian mammals by eradication of (in particular) foxes and feral cats

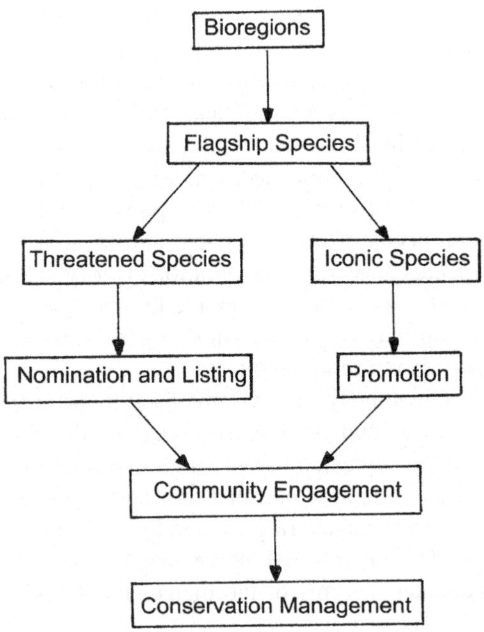

Fig. 9.1 A strategic framework for the conservation management of Australian insects and allied invertebrates from the Bioregion-based approach discussed by Taylor et al. (2018), and in which two distinct groups of 'flagship species' are included. These have rather different avenues to adoption but converge to common trajectories to conservation (after Taylor et al. 2018)

removes a key threat but also assures a predator-free fenced or island area in which risks of extinction for small mammals are markedly reduced. Woinarski et al. (2023) used the term 'havens' for these predator-free areas, and noted that the approach had wider implications for other taxa. Havens include areas from where the threatening species has been excluded, eradicated, or never occurred, sites unsuitable for the threatening species to occur, and sites far distant from the current distribution of the threatening species or where it is controlled intensively to below a threat level. Other native taxa may benefit from those changed circumstances.

9.4 Insects in Ecological Communities

The suggested wider representation of insect life across Australia's major ecosystems reflects transition to the 'coarse filter' approach to insect conservation—namely of focus on ecosystems rather than on species alone. Frameworks exist for this—Australia has a series of recognised 'biodiversity hotspots', of legally recognised 'threatened ecosystems', many of them protected areas such as national parks, and other reserves, each of which could be a basis for systematic comparative surveys and inventories of resident insects. For each, estimates of richness, endemism, and other faunal features would do much to augment their perceived values for insects—as already evident from the surveys that exist. The transition from 'species' to 'diversity' is here tempered by an intermediate step of focus on selected species as ambassadors for wider diversity, through planned inclusions of species from regionally significant ecosystems across the country. Taylor et al. (2018) considered that focus on the IBRA regions was useful, because the other units, such as designated threatened ecological communities, were variously too broad or too fine across the landscape to enable clear nominations of representative insect species, notwithstanding that individual species may be listed from them. Their suggested strategy thus includes the major ecosystem diversity across the country as incorporated into the IBRA categories. Recognition of suitable flagship species for each may become a protracted process for some IBRA bioregions, but does not obscure the regions' significance and need to conserve their insect fauna. The strategy could lead to a very broad approach of accepting notable insect taxa as umbrella taxa for each of these major regional entities.

The process of focusing on insect diversity is aided by formal recognition of 'threatened ecological communities' and 'key threatening processes' under the EPBC Act 1999 and several regional legislations in Australia. Both aspects broaden the scope of practical conservation obligations to include the environments in which species live, and impacts on them. Those communities include many of prime importance for insects, and their listing implies threshold definitions for condition, species richness, and species composition.

Most of the 84 designated threatened ecological communities (at early 2023) are terrestrial, and most are defined primarily on floristic data. They collectively embrace (as of August 2023) particular woodlands, grasslands, forests, shrublands, wetlands, marine areas, ground springs, and cave communities. From a management perspective, focus on communities is far more holistic than concentrating on species alone because it is inclusive of the complex suites of interactions, processes, and species-specific responses that abound in diverse assemblage and sustain ecosystem services—amongst which insects are often the most diverse players. Definition of the communities on floristic characters and associated landforms provides relatively stable descriptive features (Fraser et al. 2019), but presumptions of long-term stability may lead to overlooking and misunderstanding environmental changes that can affect ecosystem function. Saunders et al. (2021b) affirmed that mandating ecosystem-level protection in legislation and policy is essential to preventing biodiversity loss but, because ecosystems change continually, it is also necessary to treat them as dynamic and vulnerable—and protect them accordingly. They commented that the EPBC Act had 'so far been ineffective in protecting habitat and ecological communities'. Considering the effects of climate change on threatened temperate grassland communities in Tasmania, Harris et al. (2015) noted that attempts to 'maintain the status quo in terms of species composition may be futile in many regions of the world', and that successful conservation must incorporate ecological dynamics rather than focus wholly on a static current condition. Effective conservation management depends on the ability to make adaptive changes to incorporate these changed dynamics.

Need for listing of a threatened ecological community under the EPBC Act has to be justified against a series of criteria, for which available information may not be sufficient to include all. As an example, the Natural Temperate Grassland of the Victorian Volcanic Plain (in which the Golden sun-moth occurs) was accepted because (1) the decline in geographic distribution is very severe; (2) its geographic distribution is very restricted and coupled with demonstrable threats; (3) reduction in community integrity is very severe and regeneration is unlikely to occur within the immediate future; and (4) the rate of continuing detrimental change is substantial. The first three of these gave it a status of 'critically endangered, and the last ranked it as 'vulnerable'.

Many such geographically defined communities can be delimited clearly. This is not always the case. In Victoria the entity termed 'Butterfly Community no 1' listed under the Flora and Fauna Guarantee Act in 1989 was founded on Mt. Piper, near Broadford (Jelinek et al. 1994; Britton et al. 1995) but defined on the presence of particular rare butterflies, leading to much debate over where the community was bounded (New 2011). It was then the only place in Victoria where two rare myrmecophilous species of *Acrodipsas* butterflies (Ant-blues, Lycaenidae) were known to co-occur. The community was delimited by their presence within a substantial assemblage of other butterflies and recognition that Mt. Piper was threatened. The community is defined primarily on the suite of resident 'core' species of Lepidoptera (including *Acrodipsas*), with recognition that variations occur through vagrant and migrant species, some using the peak for 'hilltopping' (a behaviour whereby individuals aggregate on a hilltop or other high topographical point and increase their opportunities for mating encounters through increased numbers and exposure). Thus, although the intention was to emphasise the butterfly community—so that any other place where the two Ant-blue species are resident would be included, many people consider 'Butterfly Community no 1' to be synonymous with 'Mt Piper' as a single locality.

The idea of 'community' also implicitly considered the extent of biodiversity, endorsing the principle of conservation efficiency (Joseph et al. 2009) that if site triage is needed the sites with higher richness or greater representation of 'listed' species (of all taxa) should be preferred for protection.

Preparation of a management plan or recovery plan is obliged for most listed threatened communities in Australia, unless specific grounds for not doing so are stated. Recognition of key threatening processes and approaches to abating them are integral to any such plan. These will vary in detail across different ecosystems and sites within each region, reflecting the relative impacts of the drivers discussed in Chap. 3, but will demand the widest possible cooperation and liaisons with stakeholders and long-term planning. Thus, the 'Recovery Plan for Butterfly Community No 1' (New 1998) had three broad objectives: (1) to alleviate and manage current and future threats likely to lead to decline or local extinction of the constituent species; (2) to foster increase in population size of several (listed in plan) species; and (3) to maintain butterfly diversity. Recovery criteria were listed, and a series of 13 actions listed and discussed.

Conservation of insect diversity also spans attention to assemblages in anthropogenic environments, notably in extensively modified urban and agricultural areas (Chap. 3), and also to maintaining key ecological processes such as pollination. Despite loss of pollinators in Australia being less extensively publicised than in the northern hemisphere, Pyke et al. (2023) recommended that immediate actions should be taken to address its own pollination crisis (Table 9.5).

Other concerned entomologists have also nominated priorities for the research needed to guide insect conservation policy. Thus, the agenda proposed by Saunders et al. (2021a) for wide-ranging activities in the face of increasing fire risk demonstrates well the shortcomings of current capability and understanding.

Table 9.5 The 'sensible steps' suggested by Pyke et al. (2023) to address and mitigate Australia's 'pollination crisis' (used under Creative Commons Attribution licence CC-BY-4.0

1	Increase taxonomic work on suspected plant pollinators
2	Protect pollinator populations threatened with extinction
3	Establish long-term monitoring of plant–pollinator relationships
4	Incorporate pollinators into sustainable agriculture
5	Restrict use of various pesticides
6	Adopt an integrated pest and pollinator management approach
7	Develop partnerships with First Nations peoples for research, conservation, and management of plants and their pollinators

9.5 An Agenda

The following themes, some of them with applications far beyond insects, and presented as an annotated list, draw from much of the earlier discussion (such as the scheme for fireflies outlined on p. 111, and several parallel plans), founded in concerns for the future of Australia's insect fauna and expressed—some of them repeatedly—over recent decades. They encompass several major topics, ranging from basic science and conservation management to administration and policy. It is important to acknowledge the 'precautionary principle' that there is considerable doubt over the extent of insect diversity in many key habitats and the extent to which ecologically specialised insects may continue to depend on these habitats or are narrow-range endemics vulnerable to environmental changes. That doubt does not obviate conservation need. Calls for changes in Australian conservation policy to more specifically include invertebrates are not new, and calls for increased support for education and citizen science are also widespread, as a central need to facilitate interest and progress and improve basic knowledge. However, increased leadership by government agencies and effective coordination across regional legislations can emphasise significance of insects and ensure that sufficient review and adaptive management occurs for progress to be constructive. Many of the topics listed below were also raised by Yen and Butcher (1997), and remain as important and urgent now as when they were first expressed. Goulson (2021) noted a series of 'Actions for Everyone' based on the United Kingdom priorities, under four main categories: encouraging environmental awareness, greening our urban areas, transforming our food system, and improving protection of rare insects and habitats. All resonate strongly for Australia.

The themes listed below move from an initial suite of general needs and considerations (most with relevance far beyond conservation in increasing fundamental understanding of the insect fauna) to the more particular demands for practical support and process in different environments, and the decisions that may have to be taken. The themes suggested, some posed as questions, are a personal compilation for guidance and do not reflect the formal priorities of any authority or organisation in Australia; they are intended to demonstrate the variety of factors implicit in developing effective insect conservation and to stimulate further thought on how they may be developed. They are also idealistic, many of them deceptively complex, and unlikely to be fulfilled completely.

Defining the need: basic knowledge

1. Augmenting knowledge of the richness and distribution of Australia's insect diversity
    i. Continue and accelerate survey and documentation of the insects of National Parks and other key reserves, and of major terrestrial and inland water habitats throughout Australia.
    ii. Assess patterns of insect richness, endemism, and biodiversity 'hotspots' to suggest conservation importance and priority of each area surveyed.

iii. Ensure that data from both the above is readily accessible to interested parties. (It may occasionally be necessary to more closely restrict information for species of especially high significance or vulnerability and which may be sought for commercial purposes, and for which an individual case for greater security of data might be warranted).
iv. Where possible, plan to monitor changes in insect faunas by repeat sampling of key sites (such as biodiversity hotspots and unique habitat patches) at intervals (perhaps, each decade to correspond with the Red Data list process), together with documenting changes in site condition (including photographic records).
v. Develop an agreed, standardised, and coordinated sampling protocol for the above, to augment more opportunistic and widespread sampling/collecting (wherever possible, use methods that can be deployed by non-specialists with minimal costs and effort, and minimise take of bycatch; consider and encourage use of non-lethal methods).

2. Increasing national ability to recognise and name Australia's insect species

   i. Recognise and acknowledge that species-level identification is a core of evaluating insect biodiversity.
   ii. Increase taxonomic capability to accelerate collection, recognition, and description of Australia's insect fauna.
   iii. Increase production and availability of recognition aids (print or web-based) to allow non-specialists to identify insects—at least to family or genus level, and also to show where 'simple' identification is unreliable or unwise.
   iv. Increase institutional capacity to house and curate comprehensive voucher material from insect surveys.
   v. Create pathways for individual institution insect holdings to be coordinated and become an effectively integrated and accessible reference resource, probably including increased digitisation of specimens and records.
   vi. Facilitate access to taxonomic information by conservation practitioners needing to identify key focal insect species (perhaps including a regularly updated directory of taxonomists and reference sources).

3. Setting priorities amongst taxa for conservation attention and for taxonomic clarification

   i. Determine (perhaps by expert consensus) the taxonomic groups (order, family, genus) that have greatest significance or value in conservation.
   ii. Represent those priorities, with reasoning, to the relevant (Commonwealth or State/Territory) range agencies and taxonomic institutions.
   iii. Promote those taxa for accelerated taxonomic and biological attention; seek funding for research on those taxa.

4. Determining the conservation status of members of those selected groups

   i. Assess the selected taxa using the IUCN Red List categories; if further work is needed to provide conservation assessment for selected taxa, attempt to arrange for that to be undertaken. Include provision for those same taxa to be re-assessed at intervals (perhaps, a decade, in line with IUCN Red List recommendation).

ii. Determine/select which of these taxa (or components of them) should receive conservation priority, on transparent grounds based on the fullest information available, and evaluation against stated criteria (such as the IUCN Red List criteria).

5. Assessing the extent, features, and security of the major terrestrial and freshwater habitats frequented or needed by those taxa

    i. Determine and list the most important insect habitats and sites across all terrestrial and inland water ecosystems, for both richness and presence of unique or other 'notable' species, including local endemic taxa.
    ii. Assess the representation of insect habitats and notable species in current protected areas, and note the key ecosystems for which greater formal protection is warranted.
    iii. Seek that additional protection to increase the extent of insect representation in protected areas.
    iv. Where necessary, improve security of tenure and increase threat prevention/mitigation for the areas selected, whether individual sites or more broadly designated habitats.
    v. Again where necessary, develop and promote coordinated management across different land tenures needed to secure the conservation habitat/site target(s).

6. Evaluating major threats and how these might be mitigated

    i. Appraise causes of concern, notably past, continuing, and likely future threats for conservation priority insect taxa and habitats.
    ii. Where possible, document and evaluate impacts and ascertain the patterns of the most significant threats.
    iii. Determine how threat reduction and conservation enhancement may be achieved.
    iv. Formulate detailed and costed plans/protocols for this to occur, with activities prioritised.

7. Establishing the principles of conservation planning for insect species, assemblages, and habitats

    i. Recognise that insect conservation extends beyond individual species to encompass the wider diversity of assemblages, habitats, and communities, and from local to landscape/national scales, and that each may need an individual conservation programme superimposing individual detail on a more general conservation scheme.
    ii. Use the best available information to develop a comprehensive conservation plan for each priority entity, and indicate the entities with most urgent need.
    iii. Ensure that the details and rationale of each insect conservation programme are recorded and available for future reference and wide consultation

Developing practical conservation

1. Selecting the projects to be undertaken, with grounds for justification
   i. Evaluate the importance and relevance of each insect conservation project proposed for the taxa/habitat/area and assign a priority order on defensible grounds (some of the decisions to be taken are noted under 'Key foci and decisions', below).
2. Defining the scope and priority aims of a project
   i. Each project should have one or more clearly stated objectives, with priorities, targets, review intervals, and monitoring strategies stated in as much detail as possible.
   ii. Scientific and agency leadership, with facility to seek advice elsewhere as needed, should be clearly defined and agreed by all parties (stakeholders, below) involved.
   iii. Sampling/assessment methods should also be defined, with alternatives when anticipated, and due attention to use of non-lethal methods
3. Enroling and educating stakeholders
   i. All relevant stakeholders should be involved from the commencement of a project, and this base should be augmented as necessary as the project proceeds.
   ii. Project information and proposals should be apparent to all stakeholders and differences of opinion harmonised for the best common good. Continuing communication is essential, with the stakeholder representatives forming a 'working group' to oversee the project through all stages.
   iii. Effective agency leadership can help to assure continuity and monitoring of management outcomes.
4. Assessing and assuring logistic resources
   i. The financial, logistic, scientific, and other resources needed for the project should be defined by prior planning and, as far as possible, assured.
   ii. Ensure that the project has effective management for deployment of resources, and for seeking additional resources (including finance) as needs arise.
   iii. Formulate a contingency plan for possible shortage of resources as the project proceeds (for example to define the aspects in which economies may be possible and those that are 'core' and must continue).
5. Setting criteria for success or change
   i. A project should not be allowed to become indefinite or 'open-ended', but have clear aims and criteria for progressive assessment at intervals, using defined SMART criteria.[1]

[1] SMART is an acronym for 'specific', 'measurable', 'achievable', 'relevant', 'time-bound' as a desirable scheme of objectives for project assessment.

ii. Review outcomes of assessment as a guide to progress and, if necessary, modify progress through adaptive management.

6. Planning for action, monitoring, and review, with options for later activities through additional or otherwise adaptive management

    i. Ensure that the plan is sufficiently flexible to allow/enable modifications as adaptive management if these are indicated from monitoring or changed circumstances.
    ii. Changes to the original plan should be agreed amongst all stakeholders, if necessary after an independent review.

Key foci and decisions

1. Single-species focus or wider insect diversity?

    The persistent value of individual threatened species as tangible 'ambassadors' for insect conservation endorses that their conservation is a cornerstone of any wider conservation interest, and a signal of value of particular sites, resources, or habitat types, as well as garnering public interest. Wider insect diversity may not be conserved through this approach, other than as a byproduct of habitat conservation, and it may be recommended to use limited logistic resources to focus on diversity, perhaps through site/habitat protection rather than commit to expensive single-species programmes. Notable species amongst that diversity may become important foci for wider advocacy, and selected for individual attention.

    i. Are species or populations of particular significance (such as only known population, unusual or isolated taxonomic lineage, outliers from main range, historical interest [such as type localities], cultural significance) or taxa categorised as 'critically endangered' amongst candidates for attention?
    ii. Are unusual habitat types (such as vegetation, waterbodies, topography, isolated remnants) or historical or cultural sites amongst candidates for attention?
    iii. Promote surveys of the insects of all designated 'threatened ecological communities', with particular attention to any designated focal taxa for each community.
    iv. In all such surveys, consider use of non-lethal sampling methods for assessment, and ensure responsible deposition/archiving of all bycatch and target material obtained.

2. Listed species and /or others?

    Under most Australian policies, formally listed species are accorded conservation priority and in principle gain access to the limited available funding. Expectation is widespread that these should indeed be the priority. The large numbers of insects that are <u>not</u> listed leads to concerns for some, and (perhaps especially for potential local flagships) drives conservation concerns—perhaps accompanied by moves to nominate that species for listing. A case for conserving non-listed Australian insect species can often be convincing, and justified by new discoveries and appraisal.

i. Continue to appraise the status and needs of all insect species signaled as of conservation concern.
ii. Solicit suggestions/nominations of such species from under-represented habitats, ecosystems, and Australian regions, and define their significance for conservation.
iii. Produce a unified nomination/listing process across all Australian jurisdictions; facilitate rapid investigation and processing of nominations.

3. Major focus/need on urban and periurban environments

The rapid urban developments in Australia, projected to accelerate to accommodate increasing human population, render both loss of periurban areas and urban open spaces critical themes in insect conservation. Preservation of remnant patches of vegetation and waterbodies and restoration of vegetation (largely by removal of alien plants and replanting with local flora) are key themes, together with promoting connectivity (9, below) and threat management (8).

i. Involve entomologists, town planners, and landscape architects in planning of urban developments, with attention to insect conservation in all such exercises and without compromising the appeal of towns and cities to people.
ii. Encourage use of the principles of Biodiversity Sensitive Urban Design (p. 133).
iii. Preserve remnant and other significant vegetation and waterbodies and protect them from despoliation.
iv. Promote restoration of degraded public areas, and protect them from further threat; plan to revegetate key areas, and to include constructed forests and wetlands (as well as natural remnants) in new suburban developments.
v. Promote construction of green roofs and walls as additional insect habitat in built-up areas and in new developments.
vi. Plan to reduce isolation of open spaces and promote connectivity by linear features (such as vegetated roadsides) or other unbuilt areas that can function as corridors or stepping stones for insects.
vii. Institute and promote management to reduce the negative impacts of urban environments on insects, notably by habitat loss, and pollution (including light and noise pollution).
viii. Urge reduction or cessation of pesticide usage by management authorities and individuals.
ix. Encourage methods to enhance insect resources in urban areas (as examples, to maintain leaf litter and decaying wood, areas of exposed soil, richness of native flora, and vegetation structural variety), and avoid measures (such as over-sanitation, excessive pesticide use, and frequent mowing) that may deplete those resources.

x. Emphasise the importance of home gardens as insect habitats, and how their roles may be enhanced.
xi. Provide basic information on insect roles and importance in urban ecosystems to governing authorities, community groups, and individuals, together with outline of practical conservation measures.

4. Major focus/need on agricultural environments

The major and extensive transformations from agriculture, both in extent and intensification, continue to remove insect habitats in many parts of the world, and link strongly with other drivers of insect declines, such as chemical pesticides and fertilisers.

i. Preserve remnants of natural vegetation and water bodies in agricultural areas.
ii. Promote heterogeneity through measures such as increasing native vegetation in field margins, intercropping, reduced field sizes, and other measures to de-intensify agriculture.
iii. Promote restoration of suitable native vegetation to provide overall increase in insect habitats, and increasing connectivity.
iv. Promote values of natural vegetation in agricultural areas as reservoirs of natural enemies valuable in pest management. Recognise also that such patches can be reservoirs for pests and be deemed undesirable by affected farm producers.
v. Reduce and focus use of chemical pesticides (insecticides, herbicides, fungicides) and reduce non-target impacts. Wherever possible, promote use of non-polluting pest management methods and foster measures to sustain local populations of natural enemies of agricultural pests.

5. Major focus/need on non-agricultural rural environments

The matrices between urban and agricultural areas form a distinct suite of insect habitats in which disturbance levels are typically lower and leaving the areas ‘more natural’.

i. Key habitats include native grasslands, woodland and forest, waterbodies, and more specialised habitats such as caves and aquifers; each may need specific conservation measures and measures to reduce isolation.
ii. Where necessary, preserve key remnants of important insect habitats in this wider landscape and promote their values in connectivity.

6. Major focus/need on protected areas

The wide assumption that ‘all is safe’ simply by being in a protected area is unrealistic; for insects management to restore or control successional change, to prevent impacts from alien species and pollution, and to ensure critical resources is usually necessary.

i. Protected areas provide venues for active conservation management of individual insect species and wider diversity.

- ii. Threat management and habitat enhancement for key threatened insect taxa should be incorporated into wider management of protected areas.
- iii. Increase awareness of insect diversity and needs in management of protected areas, and how management might be modified to increase insect conservation.

7. Landscape-wide needs

All the above largely focus on defined sites or areas, with little formal regard to the intervening areas or 'matrix'. Many important insect species and habitats extend beyond the areas circumscribed above, with grasslands, savanna, woodlands, forests, wetlands, inland waters, and others all distributed in the wider landscape.

- i. Insect conservation needs arise across all areas that insects occupy, including those in the wider and non-protected landscape or waterscape.
- ii. Seek additional protection for any areas/ecosystems/vegetation associations known to be under-represented in the existing protected areas system.

8. Threat evaluation, prevention, and amelioration

Full definition of major threats and the changes they have made in the past, are currently making, or are projected to make in the future determines the scope of a project to prevent and reduce threats and restore changed conditions. The overriding influences of climate change should also be considered.

- i. Appraise the reasons for the project in terms of threat variety and impacts to the focal species or environment.
- ii. Determine the most significant threats and how their impacts can be prevented, reduced, and/or remediated, by management, legislation, or other means. Recognise that some threats are clearly amenable to alleviative measures whilst others (notably climate changes) are less so.
- iii. Ensure that proposed management methods and regimes (such as frequency and intensity of mowing or burning; pesticide applications) do not exacerbate threats, and are as non-damaging as possible in not raising other conservation problems.
- iv. Ensure that the project plan includes full discussion of threat management for the focal insect(s) or habitat(s).

9. Connectivity and habitat integrity

Isolation of habitat patches is an important threat in also isolating populations of their resident insects, possibly leading to genetic degeneration and otherwise increasing vulnerability. Strategies to prevent isolation and restore functional connectivity between patches are a core conservation activity. For most insect species, population structure is unknown, but any known or suspected metapopulation structure endorses the need for and value of small habitat patches and movement between these.

i. Wherever possible, appraise opportunities to establish or retain 'stepping-stone' habitat patches, however small, to facilitate insect movement between major habitat patches.
ii. Emphasise the importance of more direct linear 'corridor habitats' that can also constitute primary or core habitat with appropriate management, and may be important in countering local extirpations from climate change.
iii. Enrich such areas with local flora, including known key resources for notable local insects; augment vegetation structure; and remove invasive alien weeds and any other undesirable invasive animals and plants.
iv. Road verges and medians may be of particular significance; their natural vegetation should be conserved and enhanced wherever possible, and the areas protected from dumping, pollution, and loss.
v. If translocations or other releases of insects are anticipated, ensure that receptor sites are secure and suitable and that the exercise can be monitored and all details recorded.

Advocacy, education, and practice

1. Conveying the importance of insect conservation

   i. Enlist advice from the conservation committees and interest groups of the national (Australian Entomological Society) and regional entomological societies on key points to include in a campaign to promote insect worth in Australia, involving education and communication at all levels.

2. Increasing school and university education in entomology
   The decline of teaching of entomology in Australian universities and of natural history in schools, and wider replacement of field courses by web-based and other laboratory-based exercises has eroded the recruitment of insect enthusiasts and professional entomologists—in conjunction with reduced employment opportunities in entomology. The Australian Entomological Society's Education Committee continues to advocate for increased entomological content in secondary and tertiary education.
3. Increasing liaison with government and local government agencies
   Linking with 'Policy, no 2', below, a clear avenue for communication and coordination on insect conservation needs and projects with designated personnel in a relevant environmental agency that can help support and coordinate the project is highly desirable. Such a contact, in conjunction with the project coordinator/group, can provide 'official' recognition and advocacy, and facilitate access to funding and other support.
4. Fostering citizen science
   Wide community support is a key element of insect conservation, augmenting the limited expertise and availability of 'professional scientist' inputs. Participation of both individuals and community groups as 'citizen scientists' in all aspects of survey, monitoring, and management may be critical in an insect conservation programme. Some such cases have been initiated by concerned

citizens. The ten principles of citizen science formulated by the Australian Citizen Science Association should be considered in every project (p. 150).

i. For every insect conservation project, assess the resources available and work needed, and suggest and document roles in which citizen scientists could help. Recognise that citizen science support cannot be 'taken for granted' but also that it may be vital to project success.
ii. Explore possible sources of citizen science input, for example local naturalist groups, 'Friends' of particular sites, educational institutions.
iii. Enlist the help of local volunteers to support insect conservation programmes, and ensure that their aid is appreciated and acknowledged.
iv. Clear instructions and direction on methods and purpose of citizen science participation are necessary.
v. Have realistic expectations of the level and extent of contributions that can be made by volunteers, and of the need for supervision, reporting, and review.
vi. Encourage contributions of photographs of insects to iNaturalist or similar web site, and of other records as appropriate.

5. Providing guidance and quality control of records and observations

i. Recognise that many insect species are difficult to identify unambiguously, that photographs may not always suffice, and that voucher specimens may be needed for more critical examination.
ii. Provide for a central register of records and observations on significant insect species, as data that can be available in conservation programmes.

6. Promoting effective communication and coordination with all stakeholders

i. Provide wide publicity for specific insect conservation exercises, with explanation of need, objectives, and anticipated outcome.
ii. In all relevant cases, liaise with First Nations Australians to explore cultural boundaries and cooperation opportunities.

7. Overall, aim to transform perceptions of insects to establish insects as targets in conservation rather than as fortuitous 'passengers', through increased awareness of their diversity and importance, and wider consequences of their loss

i. Promote the principle that people are stewards for wildlife, including insects.
ii. Promote insects as direct and worthy subjects of conservation activities, of functional, economic, cultural, and aesthetic significance.
iii. Recognise that insect conservation may require specific targeted management, in addition to more general measures.
iv. Promote inclusion of insect displays in local environmental events, festivals, and exhibitions to help publicise their diversity and values.
v. Recognise that social media are a powerful means to influence public opinion and that 'postings' on any insect conservation project have the potential to garner support.

Policy

1. Increase profiles of insects in environmental policy at national and state/territory levels.
2. Increase international appreciation of the unique nature of Australia's insect fauna and the needs to investigate and conserve it.

   i. Australia's unique insect fauna has global significance, and support for conservation from scientists in other parts of the world may increase international profile of particular cases.
   ii. Encourage support for insect conservation activities in Australia from the global community of entomologists.

3. Encourage the permanent employment of an 'invertebrate specialist' in each state and territory environmental agency

   (This call is long-standing in Australia and is gaining traction, but needs to be reinforced at every opportunity. At present, invertebrate expertise is very limited in some agencies, with the result that preference for conservation attention and resources is often given to taxa (mainly vertebrates) with which staff are more familiar, and insects and others bypassed for lack of confidence or knowledge).
4. Ensure effective compatibility between Commonwealth and more local legislations in listings of insect species and ensuing commitments

   Recent revision of the EPBC Act has done much to standardise the previously varied requirements and criteria for 'listing' species across Australian legislations.
5. Facilitate and streamline the 'permit process' to allow bona fide naturalists/students/non-professional entomologists/citizen scientists greater access to protected areas and to collect/survey insects there, if necessary under supervision, and with expectation of reporting findings.

   i. For any fieldwork, including capture of specimens, in protected or privately owned areas, ensure that permission for entry and the work anticipated has been obtained from the managing agency or land-owner and that conditions of the permit are followed.
   ii. Encourage responsible deposition of voucher specimens and data from such activities in a recognised institution.

6. Explore possible synergies with other conservation projects and benefits to insects from these

   i. Recognise that many management measures for insects have far wider benefits to biodiversity through preservation or restoration of natural habitats and resources.
   ii. Equally, recognise that conservation measures for vertebrates, plants, and other biota may give substantial benefits and opportunities for insect conservation, whether fortuitously (such as by greater habitat security, as in 'havens': p. 128) or through more specific management (such as promoting

heterogeneity); in many cases insects can be transformed from 'passengers' to more prominent components of conservation programmes.

iii. Explore opportunities to advance insect conservation from other conservation projects (e.g. occurrence of notable insects on sites selected for other conservation purposes, possible insect flagships that can augment other campaigns, management steps for insects that can benefit other focal taxa).

iv. Projects undertaken to preserve or rehabilitate wetlands or terrestrial vegetation as 'general conservation' also provide resources for insects; with notice, entomological advice may provide an opportunity to enhance or introduce key resources needed by locally important insects as well as enhancing wider insect diversity.

v. Integrate all complementary conservation measures towards a common objective for conservation, and ensure that monitoring and management flexibility remain sufficient to accommodate changes in logistic supply.

References

Akcakaya HR, Bennett EL, Brooks TM, Grace MK, Heath A (et al.) (2018) Quantifying species recovery and conservation success to develop an IUCN green list of species. Conserv Biol 32: 1128–1138. https://doi.org/10.1111/cobi.13112

Arnold JP, Murphy MV, Didham RK, Houston TF (2019) Rediscovery of the 'extinct' bee *Hesperocolletes douglasi* Michener, 1965 (Colettidae: Colletinae: Paracolletini) in Western Australia and first description of the female. J Threat Taxa 11:13310–13319. https://doi.org/10.11609/jott.4610.11.3.13310-13319

Braby MF (2018) Threatened species conservation of invertebrates in Australia: an overview. Austral Entomol 57:173–181. https://doi.org/10.1111/aen.12324

Britton DR, New TR, Jelinek A (1995) Rare Lepidoptera at Mount Piper, Victoria—the role of a threatened butterfly community in advancing understanding of insect conservation. J Lepidopt Soc 49:97–113

Cordero-Rivera A (2014) Behaviour and ecology of *Hemiphlebia mirabilis* (Odonata: Hemiphlebiidae). Report on work done in Victoria during November and December 2013. https://natureglenelg.org.au

Cordero-Rivera A (2016) Demographics and adult activity of *Hemiphlebia mirabilis*: a short-lived species with a huge population size (Odonata: Hemiphlebiidae). Insect Conserv Divers 9:108–117. https://doi.org/10.1111/icad.12147

Davies DAL (1985) *Hemiphlebia mirabilis* Selys: some notes on distribution and conservation status (Zygoptera: Hemiphlebiidae). Odonatologica 14:331–339

DSE (Department of Sustainability and Environment) (2009) Advisory list of threatened invertebrate fauna in Victoria—2009. East Melbourne

Fraser H, Simmonds JS, Kutt A, Maron M (2019) Systematic definition of threatened fauna communities is critical to their conservation. Divers Distrib 25:462–473

Geyle HM, Braby MF, Andren M, Beaver EP, Bell P (et al.) (2021) Butterflies on the brink; identifying the Australian butterflies (Lepidoptera) most at risk of extinction. Austral Entomol 60: 98–110

Gibson L, New TR (2007) Problems in studying populations of the golden sun-moth, *Synemon plana* (Lepidoptera: Castniidae), in south eastern Australia. J Insect Conserv 11:309–313

Goulson D (2021) Silent earth. Averting the insect apocalypse. Jonathan Cape, London

Harris RMB, Carter O, Gilfedder L, Porfirio LL, Lee G, Bindoff NL (2015) Noah's ark conservation will not preserve threatened ecological communities under climate change. PLoS One 10(4):e0124014. https://doi.org/10.1371/journal.pone.0124914

Hawking JH (1999) An evaluation of the current conservation status of Australian dragonflies. In: Ponder W, Lunney D (eds) The other 99%. The conservation and biodiversity of invertebrates. Trans R Zool Soc N S W, Mosman, pp 354–360

Hill L, Michaelis FB (1988) Conservation of insects and related wildlife. Occasional Paper No. 13, Australian National Parks and Wildlife Service, Canberra

Jelinek A, Britton DR, New TR (1994) Conservation of a threatened butterfly community at Mount Piper, Victoria. Mem Queensl Mus 36:115–120

Joseph IN, Maloney RF, Possingham HP (2009) Critical allocation of resources among threatened species; a project prioritisation protocol. Conserv Biol 23:328–338

Kearney MR, Hossain MA, Sinclair SJ, Song H (2020) Grasshopper country before and after: a resurvey of Ken Key's collecting expeditions in New South Wales, Australia, 70 years on. Austral Entomol 60:52–65. https://doi.org/10.1111/aen.12515

Kearney SG, Carwardine J, Reside A, Adams VM, Nelson R (2021) Saving species beyond the protected area fence: threats must be managed across multiple land tenure types to secure Australia's endangered species. Conserv Sci Pract 4:e617. https://doi.org/10.1111/csp2.617

Key KHL (1978) The conservation status of Australia's insect fauna. Occasional Paper No.1 Australian National Parks and Wildlife Service, Canberra

Kirkpatrick J, McDougall K, Hyde M (1995) Australia's most threatened ecosystem: the southeastern lowland native grasslands. Surrey Beatty and Sons, Chipping Norton

Legge S, Rumpft L, Garnett ST, Woinarski JCZ (2023) Loss of terrestrial biodiversity in Australia: magnitude, causation, and response. Science 381:622–631

Lester PJ, Brown SDJ, Edwards ED, Holwell GI, Pawson SM (et al.) (2014) Critical issues facing New Zealand entomology. N Z Entomol 37:1–13. https://doi.org/10.1080/00779962.2014.861789

Moir ML (2021) Coextinction of *Pseudococcus markharveyi* (Hemiptera: Pseudococcidae): a case study in the modern insect extinction crisis. Austral Entomol 60:89–97. https://doi.org/10.1111/aen.12506

Moulds MS (1999) The history of butterfly research and collecting. In: Kitching RL, Scheermeyer E, Jones RE, Pierce NE (eds) Biology of Australian butterflies. CSIRO Publishing, Melbourne, pp 1–24

New TR (1998) Recovery Plan for 'Butterfly Community No.1', a threatened butterfly community at Mt Piper in central Victoria. Report to Department of Natural Resources and Environment, Victoria

New TR (2009) Insect species conservation. Cambridge University Press, Cambridge

New TR (2011) Butterfly conservation in South-Eastern Australia: progress and prospects. Springer, Dordrecht

New TR (2018) Forests and insect conservation in Australia. Springer, Cham

New TR (2019) Insect conservation and Australia's grasslands. Springer, Cham

New TR (2020) Insect conservation and Australia's inland waters. Springer, Cham

New TR, Sands DPA (2003) The listing and de-listing of invertebrate species for conservation in Australia. J Insect Conserv 7:199–205. https://doi.org/10.1023/B:JICO.0000020955.94081.84

New TR, Yen AL (2012) Insect conservation in Australia. In: New TR (ed) Insect conservation: past, present and prospects. Springer, Dordrecht, pp 193–212

Pyke GH, Prendergast KS, Ren Z-X (2023) Pollination crisis Down-Under: has Australasia dodged the bullet? Ecol Evol 13:e106.39. https://doi.org/10.1002/ece3.10639

Reid C, Hutton I (2019) Citizen science and the art of discovery: new records of large Coleoptera from Lord Howe Island, July 2018. Tech Rep Aust Mus 28:1–16

Reid C, Hutton I, Thompson S (2020) The citizen science survey of large Coleoptera on Lord Howe Island, August 2019. Tech Rep Aust Mus 31:1–15

Richter A, Weinhold D, Robertson G, Young M, Edwards T (et al.) (2012) More than just an empty case: a non invasive technique for monitoring the Australian critically endangered golden sun moth, *Synemon plana* (Lepidoptera: Castniidae). J Insect Conserv 17: 529–536. https://doi.org/10.1007/s10841-012-9537-5

Sands DPA, New TR (2002) The action plan for Australian butterflies. Environment Australia, Canberra

Sands DPA, New TR (2013) Conservation of the Richmond birdwing butterfly in Australia. Springer, Dordrecht

Sant GJ, New TR (1988) The biology and conservation of *Hemiphlebia mirabilis* Selys (Odonata, Hemiphlebiidae) in southern Victoria. Arthur Rylah Institute for Environmental Research, Tech Rept ser No 82, Melbourrne

Saunders ME, Barton PS, Bickerstaff JRM, Frost L, Latty T (et al.) (2021a) Limited understanding of bushfire impacts on Australian invertebrates. Insect Conserv Divers 14(3), 285–293 https://doi.org/10.1111/icad.12493

Saunders ME, Bower DS, Mika S, Hunter JT (2021b) Condition thresholds in Australia's threatened ecological community listings hinder conservation of dynamic ecosystems. Pac Conserv Biol 27:221–230

Taylor GS, Braby MF, Moir ML, Harvey MS, Sands DPA (et al.) (2018) Strategic national approach for improving the conservation management of insects and allied invertebrates in Australia. Austral Entomol 57: 124–149. https://doi.org/10.1111/aen.12343

Upton MS (1997) A rich and diverse fauna: the history of the Australian National Insect Collection 1926–1991. CSIRO Publishing, Melbourne

Ward MS, Simmonds JS, Reside AE, Watson JEM, Rhodes JR (et al.) (2019) Lots of loss with little scrutiny: the attrition of habitat critical for threatened species in Australia. Conserv Sci Pract 1, 11:e117. https://doi.org/10.1111/csp2.117

Woinarski JCZ, Chapple DG, Garnett ST, Legge SM, Lintermans M, Scheele BC (2023) Few havens for threatened Australian animal taxa that are highly susceptible to introduced and problematical native species. Biodivers Conserv 33(1):1–27. https://doi.org/10.1007/s10531-023-02750-4

Yen A, Butcher RJ (1997) An overview of the conservation of non-marine invertebrates in Australia. Environment Australia, Canberra

Appendix: Recognising the Orders of Insects Important in Conservation Practice

Introduction: Overview

Many of the insect orders in Australia have scarcely intruded into the accounts in earlier chapters and currently have no practical role in insect conservation—despite awareness that 'they are there' and 'they are vulnerable'. As elsewhere in the world, the insects noted for conservation interest mostly belong to the more iconic groups noticed by people or which are known to have practical values as monitoring tools. This does not mean that all members of the neglected orders are secure—but simply that we do not know enough to evaluate their needs and vulnerability: in principle there is little doubt that the impacts of habitat loss, alien species, pollution, and climate change have far wider impacts on invertebrate life across all their terrestrial and aquatic environments than have yet been documented, but the chances of establishing this from hard evidence are low. Many insect orders receive little attention in Australia and are essentially 'orphaned' through lack of specialist availability. However, for any of these 'neglected' orders, advocacy for particular species can arise at any time as a specialist recognises a concern or need for conservation attention to a localised, rare, or phylogenetically important species, especially if these depend on a restricted or vulnerable habitat. Thus, two locally endemic species of silverfish (Thysanura) found in caves were signalled because of cave flooding and tourist disturbance, respectively (Hill and Michaelis 1988), and remain the only Australian Apterygota considered specifically for conservation need.

For most of the more advanced orders (the 'winged insects', Pterygota) on which information is available, some—in some cases, many—species are known from single localities, often from single specimens, and commonly have not been seen for decades or more since they were first noted. Inference that some species may be extinct is difficult to refute—with targeted searches often <u>not</u> leading to rediscovery, and other investigations confirming the restricted abundance and distribution of the target taxa. Any such poorly known species may gain conservation priority if its

T. R. New, *Insect Conservation in Australia: Why and How*,
https://doi.org/10.1007/978-3-031-66631-5

situation raises specialist interest—but the pursuit of practical conservation for such invertebrates can only rarely be based on sound evidence of need. In short, conservation interest in Australian insects (largely following similar trends elsewhere) is taxonomically very uneven, with some large, ecologically significant and vulnerable taxa almost wholly ignored.

One functional dichotomy in insect life cycles has a direct bearing on approaches to conservation. Many of the orders with an incomplete metamorphosis (such as grasshoppers and true bugs) spend immature and adult stages in the same habitat, in which larvae and adults share the same resources such as foods and live in the same climatic environments. Insects with a complete metamorphosis (such as Lepidoptera, Diptera, Coleoptera) need very different conditions to support ecologically different larval and adult stages—a chewing herbivorous caterpillar needing a particular food plant is a very different animal from a sucking nectar-feeding butterfly. Conservation of the latter must thereby consider two different environments and suites of resources, both of which are necessary to sustain the species. In contrast, a grasshopper may need only one 'habitat' across its entire life cycle—in principle perhaps a simpler-sounding exercise to accomplish. Threat profiles differ across these two scenarios: considering threat abatement for one environment may be simpler than considering different threat arrays for two environments. A related dilemma arises from those insect orders (or sections of orders) with incomplete metamorphosis that have adopted a secondary aquatic life for larval stages—for which both aquatic and terrestrial conservation measures must be integrated. This latter scenario spans the primitive Palaeoptera (Ephemeroptera, Odonata), the stoneflies (Plecoptera), and some true bugs (Hemiptera). Within the advanced insects with a complete metamorphosis, only caddisflies (Trichoptera) are universally aquatic as larvae, but the habit has also evolved independently in lineages of several other orders, including flies (Diptera), beetles (Coleoptera), and moths (Lepidoptera).

Recognition and association of larval stages is well-advanced for the major 'aquatic groups' (as with aquatic lineages of higher insects such as Megaloptera and Neuroptera) in which larvae are amongst the largest and most easily recognised benthic macroinvertebrates used to appraise the state of freshwater communities and their responses to disturbance. Many biologists are far more familiar with the components of the 'EPT Index' (p. 103) in their context as indicators than as direct targets of conservation. Nevertheless, early interest in the Large Blue Lake mayfly (*Tasmanophlebia lacus-coerulei*) confined to pollution-prone alpine lakes in the Kosciuszko National Park, New South Wales, and two caddisflies known only from Lake Pedder and its environs in southwest Tasmania (and both initially believed to have been lost when the only known site they were known to occupy was flooded) show that concerns exist, in many cases accompanied by uncertainty. Plecoptera have drawn attention in the contexts of (1) small montane or forest-associated species threatened by forest clearing or other habitat disturbance (p. 37) and (2) large alpine species found in small streams (p. 60).

Odonata are by far the most understood and popular order of aquatic insects, with conservation interests spanning single species of Zygoptera (damselflies) and Anisoptera (dragonflies) to faunal composition and responses to a wide range of changes in and around waterbodies. Their values in freshwater quality assessments are well established and link strongly with conservation interests. Thus, the 'Dragonfly Biotic Index' (DBI) derived for the South African fauna (Samways and Simaika 2016) gave each species a 'DBI value' derived from three subindex themes as (1) geographical distribution; (2) IUCN Red List threat status; and (3) sensitivity to human disturbance of their habitat, with the summed scores for all species at a site indicating its relative quality. Continuing refinement of this approach has led to a continent-wide assessment for African Odonata (Vorster et al. 2020), using the 604 species for which adequate information was available, and which may be a baseline for monitoring future changes in freshwater ecosystems and showing where conservation is most needed.

Identification of most species of Australian Odonata is possible through a recent and well-illustrated handbook (Theischinger et al. 2021) that contains colour photographs of adults and illustrated keys to larval and adult stages; the book builds on a legacy of earlier works in which knowledge of the order has advanced progressively to now comprise one of the most complete and 'user-friendly' overviews of any of Australia's regional insect groups. Together with earlier conservation status evaluations of many species (Hawking 1999), and allocation of common names to all taxa, Australian Odonata are indeed accessible to non-specialist workers and are a 'flagship group' for aquatic insect conservation—as elsewhere in the world. The unique nature of the fauna is widely recognised; the tiny *Hemiphlebia mirabilis* (p. 158) attracted global interest, and the five endemic species of *Petalura* (with the New Zealand *Uropetala*, comprising the family Petaluridae), by contrast amongst the largest of all dragonflies and with larvae estimated to take up to a decade to develop, are of conservation concern as the swamps they inhabit are threatened by drainage and conversion. The Giant petaltail (*Petalura gigantea*) depends on groundwater-dependent ecosystems in its limited range in New South Wales, where a number of populations have been lost as habitats succumbed to development (Baird and Burgin 2016; discussion in New 2020).

Many of the orders of predominantly terrestrial insects with an incomplete metamorphosis have, other than for occasional allusions to 'rare' or 'unusual' species, not been considered constructively as conservation foci. Blattaria (cockroaches and termites), Mantodea (mantids), Dermaptera (earwigs), Embioptera (webspinners), Phasmatodea (stick and leaf insects), Psocodea (parasitic lice and psocids), and Thysanoptera (thrips) rarely gain mention or acknowledgement, despite their habitat limitations, vulnerabilities, and evolutionary peculiarities. Some host-specific lice, for example, may become vulnerable as their mammal or bird host numbers decline. Taxa with higher profile include the 'Giant cockroaches' (*Macropanesthia* spp.) popular as pets, and living for up to a decade, and a variety of Phasmatodea. Some of the largest of all stick insects occur in tropical Australia (p. 12). Other foci include the Giant prickly stick insect (*Extatosoma tiaratum*) also reared easily for the pet trade, and the extended conservation campaign for the Lord Howe Island stick insect (*Dryococelus australis*) with its dependence on captive breeding to save it from likely extinction (p. 48) and the complexities of its needs discussed by Wilkinson (2014). Interest in phasmids has been fostered through the Phasmid

Study Group, an international community formed in 1980 with the mission of fostering the study of phasmids worldwide. More than 400 species are listed in this Group's 'Culture List' (at October 2023) and indicates the worldwide appeal of these often spectacular insects; the group's emblem is a female *E. tiaratum*.

Orthoptera (grasshoppers, crickets, katydids) and Hemiptera ('sucking bugs', including aphids, scale insects, cicadas, leafhoppers, and their relatives, and true bugs), are the two most diverse exopterygote orders, both with numerous Australian endemic taxa. Both are predominantly herbivorous with many Hemiptera highly host-plant specific. Pastoralisation and forestry have for long been recognised as key threats to Australian Orthoptera through direct and extensive habitat loss (Rentz 1993), leaving many taxa dependent on small remnant habitat patches of native vegetation. Conservation comments have been made in support of the flightless endemic morabine grasshoppers (notably *Keyacris scurra* on grasslands in the south-east), spectacularly coloured species such as northern Leichhardt's grasshopper (*Petasida ephippigera*, depicted commonly in tourist literature from the Northern Territory), and 'oddities' such as Cooloolidae (p. 10), but the potential for Orthoptera to become conservation flagships noted by Rentz (1993) is still to be realised. Substantial radiations of several Hemiptera groups (most notably Psylloidea) in association with native flora and the less diverse and more primitive bugs such as Moss bugs (Peloridiidae, found in the ancient wet *Nothofagus* forest of south-east Australia) exemplify the numerous peculiarities across different ecosystems, all of which need further study to appraise their full diversity and detailed conservation needs, beyond knowing that dependence on native vegetation is high.

Conservation attention to the most advanced insect groups, the orders with a complete metamorphosis (Endopterygota) and so generally with more ecologically varied life cycles, is also very uneven. Some of the most diverse groups have not been evaluated in any detail, and their daunting complexity renders the task formidable and impracticable. Vast numbers of parasitoid wasps, for example, are poorly known and their host relationships still unclear—but it is reasonable to suppose that, as elsewhere, the fate of their perhaps specific hosts determines their future existence. Neglect of such taxa, together with numerous small flies, beetles, microlepidoptera, the parasitic Strepsiptera, and others for which knowledge is fragmentary, is currently inevitable. Very rare cases occur of truly exceptional species such as the recently discovered and primitive Enigma moth (*Aenigmatinea glatzella*) on Kangaroo Island (p. 165) that garner international or national interest. The moth is now known to be the host of a probably specific parasitoid wasp that oviposits in moth eggs and has been described as a new genus and species of Braconidae (Glatz et al. 2022). Parallel discoveries of novel parasitoids of little-studied insect herbivores are likely to be widespread, and co-extinctions of participants in such obligate associations—as with specific herbivore-plant dependencies—increasingly likely. The same potential exists for fleas (Siphonaptera) and their vertebrate hosts, and for which any conservation of the parasites depends directly on measures to conserve the hosts.

The aquatic Megaloptera and the related Neuroptera, together with Mecoptera, all contain narrow-range endemic and poorly studied species and have played no active roles in conservation. Of the four 'large orders', Diptera (true flies) have received least attention, other than for glowworms (*Arachnocampa* spp., p. 47) a few aquatic midges, and the bias flowing from extensive listing of Drosophilidae noted earlier (p. 166) and that emphasised the need to conserve their rainforest habitats—and also the lack of expressed concerns for the numerous other families of flies, some of them distinctive Australian radiations. Conservation measures have been concentrated on some sections of Coleoptera, Lepidoptera, and aculeate Hymenoptera. The first two are the most 'collectable' and commercially desirable insects. Concerns for collectable Coleoptera have been raised for some rare jewel beetles (Buprestidae), stag beetles (Lucanidae), longhorn beetles (Cerambycidae), and ground beetles (Carabidae), in particular, as well as over wider habitat losses and disturbance for these and other locally endemic beetles such as cave-inhabiting taxa and the aquatic Dytiscidae limited to isolated remote aquifers. Most less conspicuous beetles remain the province of specialists alone. Likewise, butterflies are by far the most conservation-attended group of Lepidoptera and, indeed, of insects. Extensive knowledge of Australian species and continuing naturalist interests in their declines and conservation, in part stimulated by preparation of a national 'Action Plan' (Sands and New 2002; see also Geyle et al. 2021 who suggested the most threatened species), has led to butterflies dominating insect conservation agendas and lists of priority species. They are followed by some moths, most notably sun-moths (Castniidae), but—as for beetles and Hymenoptera—most of the smallest species (here 'microlepidoptera') are not considered unless flagged individually by a concerned authority.

Conservation interest in Hymenoptera has the functional bases of (1) the highly diverse native bees being known widely as key pollinators of native flora and (2) a very high ant diversity that constitutes a range of 'functional groups' that have been employed widely as indicators of environmental condition and for monitoring changes such as restoration. Both groups also have species of individual conservation interest.

The iconic large and colourful Green carpenter bee (*Xylocopa aerata*) now occurs only in two widely separated areas (near Sydney and on Kangaroo Island), with habitat loss in the intervening areas responsible for losses of the bee. Further losses from wildfire on Kangaroo Island have led to a major innovative conservation programme that includes deployment of artificial nests that resemble the natural soft grasstree (*Xanthorrhoea*) stems used by the bee (Hogendoorn et al. 2021).

Australia's ants also have their 'flagships'. The Dinosaur ant (*Nothomyrmecia macrops*), regarded as a living fossil and the most primitive of all known ants, was for long the 'holy grail' of Australian myrmecology before it was rediscovered in a small area of South Australia (Taylor 1978), after not having been seen for more than 40 years. Another poorly known bull ant known only as '*Myrmecia* sp. 17' (p. 56) was listed under Victorian legislation, leading to some debate over the recognition of undescribed species in this way. The ecological significance of ants is recognised also by the intricate mutualistic associations between particular ants and lycaenid butterflies (the 'blues' and 'coppers'), so the conservation programmes for the butterflies must also cater for these obligate associates, in which species-level recognition is vital. Nevertheless, the vast majority of Hymenoptera species, as parasitoids attacking other insects, are poorly known and scarcely heeded—and difficult to study and appraise in even the best-documented faunas (Shaw and Hochberg 2001) because of lack of expertise and interest coupled with small size and highly incomplete taxonomy. Shaw (2006) suggested that parasitoid wasps may be especially vulnerable for three main reasons: (1) operating at a high trophic level; (2) many being very specialised; and (3) almost all species being very poorly known. One practical anomaly is that very rare hosts (such as some butterflies) of apparently monophagous parasitoids may be attacked by even rarer wasps that may be considered a threat and are targeted for suppression in the primary conservation programme. Whilst conservation of any such parasitoid implicitly needs conservation of its specific host, conservation of that host may need its protection from parasitoid attack.

Diagnostic and Biological Notes

These diagnostic notes are a brief synopsis of recognition features and biology of the various taxa, and are provided simply to facilitate recognition of insect orders in Australia by non-entomologists. Several orders absent from Australia are not included, together with Zoraptera, known in the region only from Christmas Island and (perhaps) unlikely to be found elsewhere in Australia. For many readers these notes will be redundant, but for those encountering insects for the first time may—with the accompanying line diagrams—help initial attempts at identification. Many of the structural features used are unambiguous, but a good magnifying glass or a stereoscopic binocular microscope will be useful for many insect identifications, simply because of their small sizes and the comparative details involved. Any such abbreviated account cannot cover all the morphological variations within large taxonomic groups, and uncertain outcomes should be checked against the fuller accounts in texts such as Naumann (1991), or by reference to the numerous informative web sites accessible by entering the order name into a search engine and from which much relevant information may be gleaned. Any entomological text may be helpful, but the key and figures produced by Harvey and Yen (1989) are a useful introduction. The illustrations here are modified slightly from those in New (1992, 1996) and indicate only the general appearance of representative members of each order, and some features of taxonomic value. The pattern of wing venation is characteristic for most winged orders, and commonly diagnostic for genera and species. Note that many illustrations are simplified by omitting wings and/or other appendages from one side of the body; the figures are not to a common scale.

I use the term ‘conservation profile’ to suggest subjectively a scale for the amount of conservation attention the group has received in Australia: ‘low’ means non-existent to very occasional notice, whilst ‘high’ implies predominant or widespread attention. The intermediate 'moderate' reflects some more sporadic or uneven attention, perhaps to only a low proportion of needy species.

Apterygota

(the two most primitive insect orders, wingless; both orders have three long apical filaments at the apex of the abdomen, and most species are about 1 cm body length)

Archaeognatha (bristletails). (Fig. A1 a). Terrestrial. Elongate, slender, with large compound eyes, meeting on top of head; thorax strongly arched.

Thysanura (silverfish). (Fig. A1 b). Terrestrial. Elongate, slender, compound eyes small or absent; thorax not arched.

(conservation profile for both orders: low)

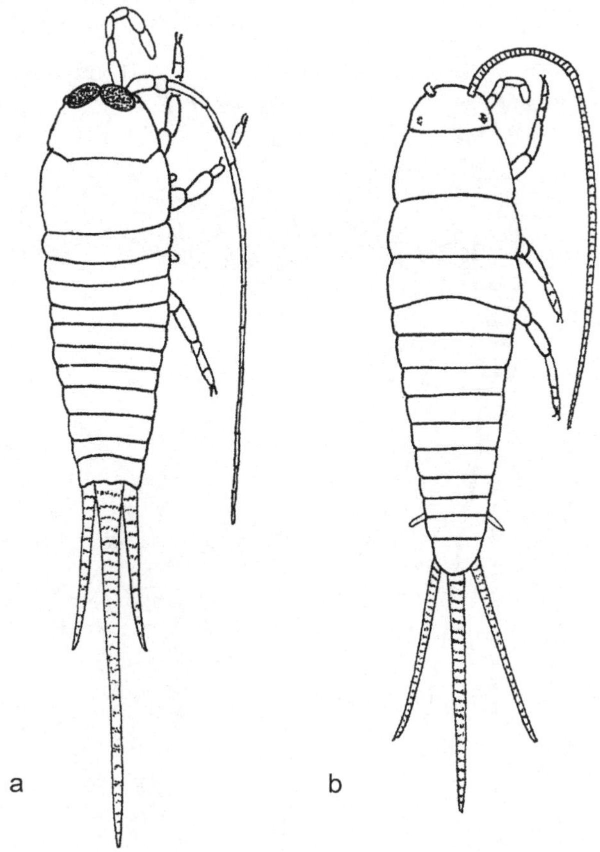

Fig. A1 General appearance of (**a**) a bristletail (Archaeognatha: *Allomachilis froggatti*) and (**b**) a silverfish (Thysanura: *Lepisma saccharina*)

Pterygota

(all the following orders are in this series, although some have secondarily lost one or both of the two pairs of wings)

Palaeoptera

(the most primitive winged insects)

Ephemeroptera (mayflies). (Fig. A2). Aquatic, with short-lived (non-feeding) and weakly flying adults rarely moving far from water. Fore wing large, hind wing reduced in size, rarely absent; abdominal apex with three (rarely, two) long

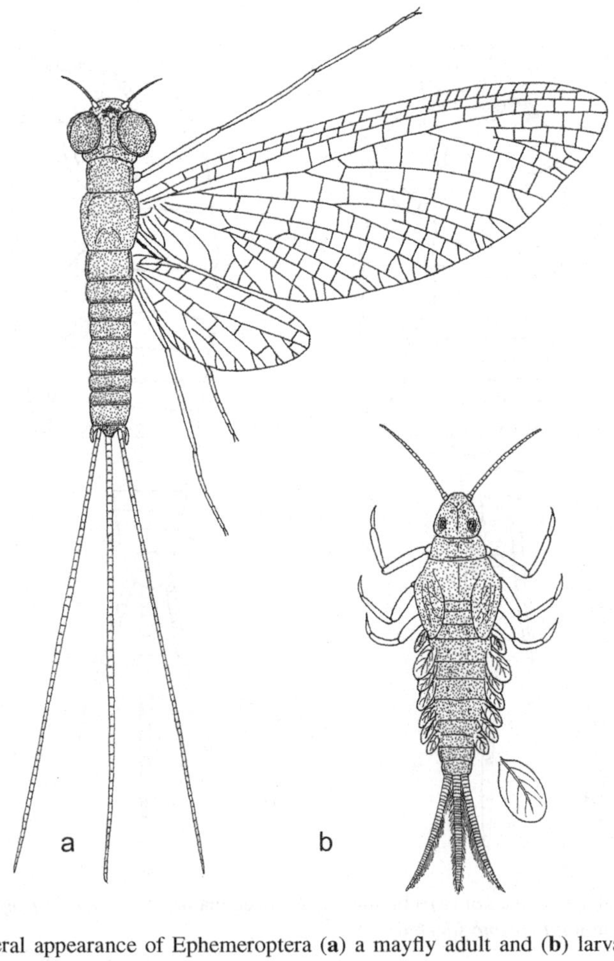

Fig. A2 General appearance of Ephemeroptera (**a**) a mayfly adult and (**b**) larva (single lateral abdominal gill inset)

segmented filaments, usually longer than body; compound eyes of male usually large. Larvae with chewing mouthparts, three apical abdominal processes, and up to eight pairs of lateral abdominal gills. The only insects with two winged instars; the last instar larva moults to a fully winged 'subadult' (subimago) which moults to the reproductive adult stage. Most species about 1–2 cm body length.

(conservation profile: high; mostly as larval constituents of indicator series [EPT])

Odonata (damselflies, dragonflies). (Fig. A3). Aquatic, active predatory larvae ('mudeyes') feed on a variety of aquatic animals, and adults capture and feed on

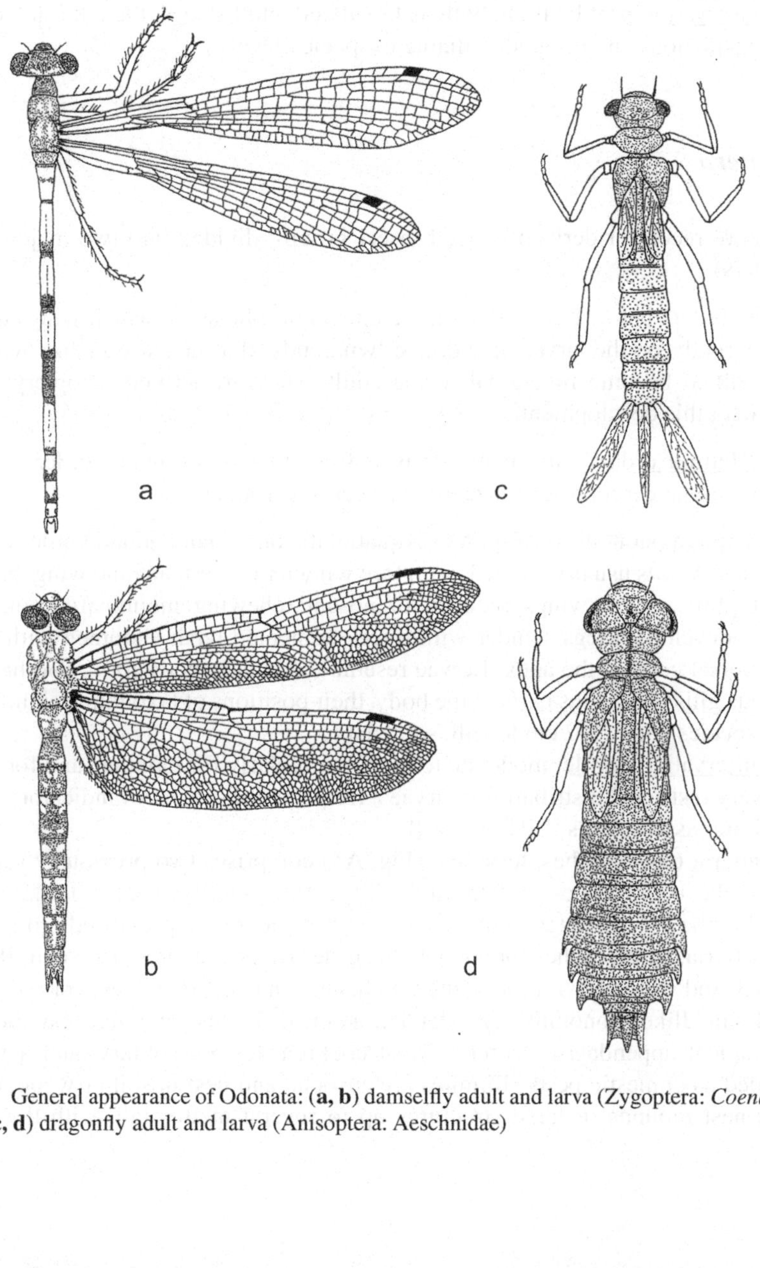

Fig. A3 General appearance of Odonata: (**a, b**) damselfly adult and larva (Zygoptera: *Coenagrion* spp.); (**c, d**) dragonfly adult and larva (Anisoptera: Aeschnidae)

small flying insects. Adults are large aerial predators, with two pairs of long wings with dense venation and very large compound eyes and long slender body; antennae short and bristle-like; legs short, often with strong spines. Larvae with chewing mouthparts capture prey on an elongate hinged labium unique to this order; eyes large. The two major series are the Zygoptera (damselflies, with slender wings held vertically when at rest; larvae generally slender, active, with three leaflike posterior 'gills'), and Anisoptera (dragonflies proper, generally large sturdy insects with wings held horizontally when at rest, fore wing narrower than hind wing; many species are brightly coloured; larvae without posterior gills, often rather short and squat, as ambush predators).

(conservation profile: high, both as larval and adult stages, the latter large, diurnal, conspicuous and often identifiable to species-level)

Neoptera

(the more recent or derived winged insect groups, divided into two major groups of orders)

A. The orders below have an incomplete metamorphosis, in which wings develop externally on the larvae as thoracic 'wingbuds' that increase in size with each moult to become functional in the adult. They are termed 'exopterygote' to reflect this development.

The 'Blattoid orders', the name derived from that of its major order, Blattaria (abdominal cerci present as paired projections from apex).

Plecoptera (stoneflies). (Fig. A4). Aquatic: the only major aquatic order of exopterygotes. Adults usually winged, with fore wing narrower than hind wing, but some taxa flightless with wings reduced or absent; chewing mouthparts, sometimes weakly developed; legs slender with three-segmented tarsi; abdomen with pair of short dorsal cerci at the apex. Larvae resemble adults in body form, and have filamentous gills on various parts of the body, their positions of taxonomic significance. Most species about 1–2 cm length, some larger.

(conservation profile: moderate to high, particularly for montane and forest taxa with very restricted distributions; larvae a frequent constituent of indicator series in freshwater assessments [EPT, p. 103])

Blattaria (cockroaches, termites) (Fig. A5) comprises two previously separated orders, the cockroaches and termites. Terrestrial. Until recently, termites were placed in the order Isoptera, but they are now considered a specialised group within the Blattaria. Cockroaches (previously the order Blattodea) are dorsoventrally compressed, and winged (with fore wing toughened, and venation dense) or wingless; a broad shieldlike pronotum; legs slender, adapted for fast running; abdomen with short apical appendages ('cerci'). Most cockroaches are solitary, and a few are regarded as domestic pests. Termites are eusocial and nest in soil or wood, or construct nest mounds or large structures up to several metres tall, with the colony

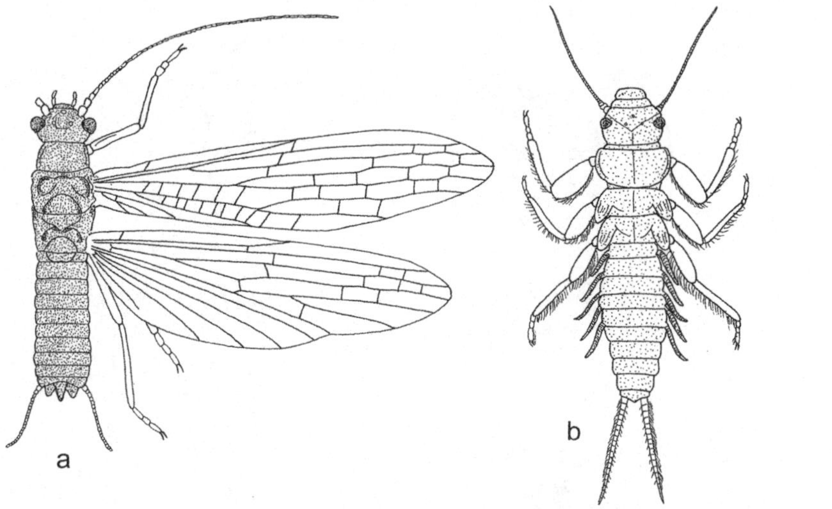

Fig. A4 General appearance of Plecoptera: (**a**) stonefly adult (Gripopterygidae) and (**b**) larva (Eustheniidae)

divided into reproductive, soldier, and worker castes; wings of reproductives are slender, and are dropped off after dispersal from the parent colony; eyes present (reproductives) or absent (others); antennae are moniliform ('beadlike'); cerci (apical abdominal projections) are very short.

(conservation profile: generally low, appreciation of termite ecological roles increases their profile over others)

Mantodea (praying mantids). (Fig. A6 a) Terrestrial. Predators, distinguished by raptorial forelegs with strong spines, other legs long and slender, and large compound eyes; head mobile; pronotum elongated; winged or wingless. Most species are well camouflaged and resemble vegetation. Most are large.

(conservation profile: generally low)

The 'Orthopteroid orders', the name derived from the predominant order Orthoptera.

Dermaptera (earwigs). (Fig. A6 b). Terrestrial. Elongate and characterised by apical cerci being forcepate; winged or wingless; when present, fore wing reduced to a small hardened structure protecting foldable large hind wing; legs short with three-segmented tarsi.

(conservation profile: generally low)

Orthoptera (grasshoppers, crickets, katydids). (Fig. A7). Terrestrial, and by far the most diverse order in this series; chewing mouthparts, directed ventrally; dorsal part of prothorax (the 'pronotum') extended laterally to form sides of the prothorax; winged or wingless; fore wings may be toughened and, with hind legs, modified for sound production; hind legs usually enlarged (jumping). Most species are herbivores, some being serious sporadic pests of pasture and/or field crops. True grasshoppers and locusts have short antennae and cerci, and may be winged or wingless;

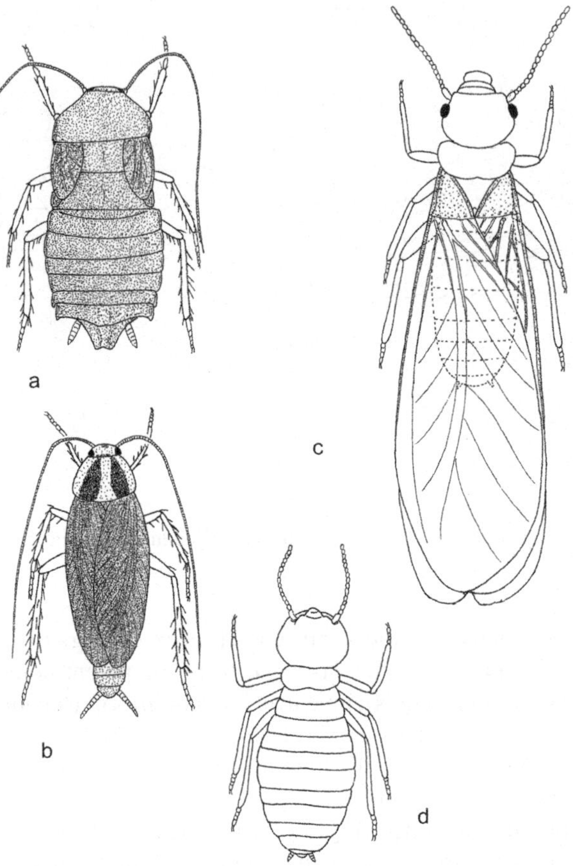

Fig. A5 Blattodea: (**a, b**) cockroaches (Blattidae, Blatellidae), (**c, d**) termites (winged reproductive, apterous worker)

taxa with long antennae include crickets (many with wings held horizontally over the abdomen and the hind wing protruding; cerci often long) and the katydids, many well-camouflaged on vegetation and with a large blade-like ovipositor. Generally large.

(conservation profile: moderate, but with some flagship taxa demanding higher attention)

Phasmatodea (stick insects, leaf insects). (Fig. A8 a). Terrestrial. Chewing herbivores; winged or wingless; fore wing short, hind wing large; antennae usually long and all legs long; mostly associated with trees and, as their common names imply, well-camouflaged by their resemblance to twigs or foliage. Some forest stick insects are amongst the largest insects known.

(conservation profile: moderate to high as a 'popular group')

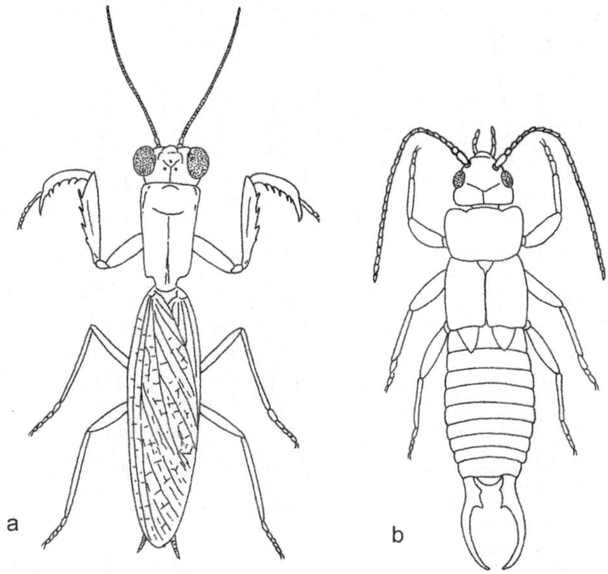

Fig. A6 General appearance of Mantodea and Dermaptera; (**a**) a mantid (*Orthodera*); (**b**) an earwig (*Forficula*)

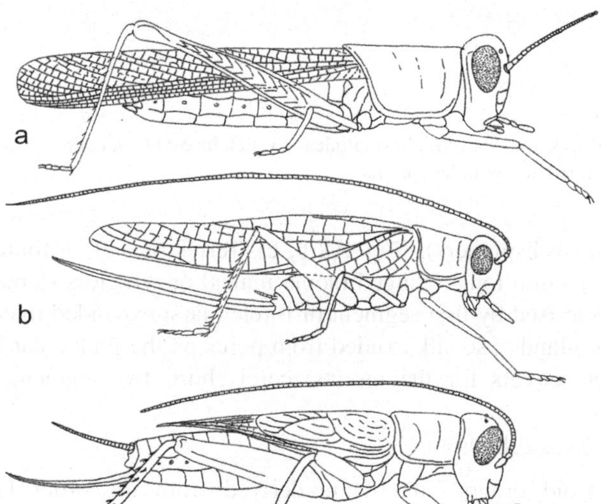

Fig. A7 Orthoptera: three major groups, in profile, as (**a**) short-horned grasshopper (Acrididae), (**b**) katydid or long-horned grasshopper (Tettigoniidae), and (**c**) cricket (Gryllidae)

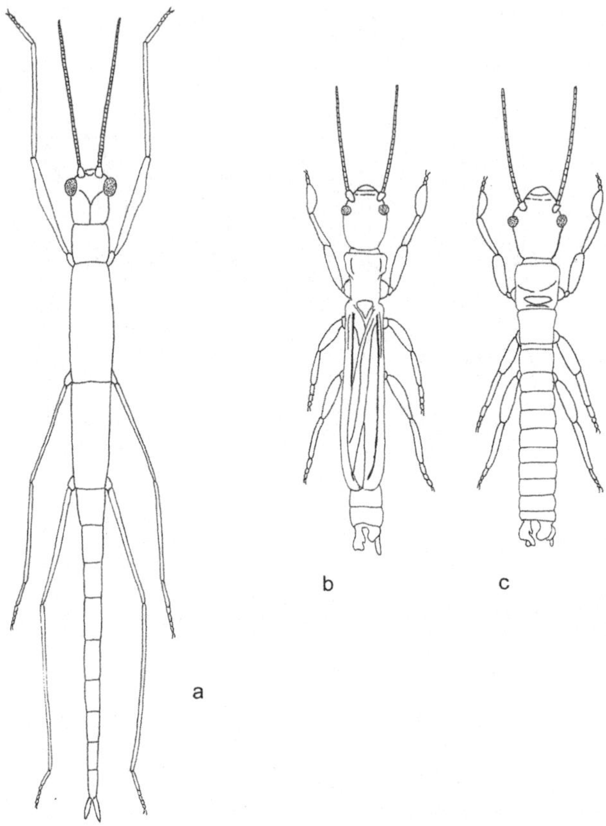

Fig. A8 General appearance of (**a**) Phasmatodea—a stick insect (*Podacanthus*); (**b, c**) Embioptera, a webspinner, winged and wingless adults

Embioptera (webspinners). (Fig. A8 b, c). Terrestrial, often found in colonies. Slender, ca 8–10 mm long. Winged (some males) or wingless (females and some males). Characterised by first segment of foreleg tarsi expanded to contain numerous small 'silk glands', so silk exuded from pores on these tarsi can construct tunnels, webs, or retreats for the insects; cerci short, two-segmented, sometimes asymmetrical.

(conservation profile: low)

The 'Hemipteroid orders', the name derived from the order Hemiptera (all lack cerci).

Psocodea (Fig. A9). Comprises two groups previously regarded as distinct orders. These are Phthiraptera (parasitic lice) and the non-parasitic barklice and booklice formerly treated as a distinct order, Psocoptera. Most parasitic lice, all wingless and only a few mm long, and many of them with specific bird or mammal hosts, are dorsoventrally flattened (cf. fleas) and with tarsi of short thickened legs modified to grasp feather plumules or hairs; antennae very short; different groups

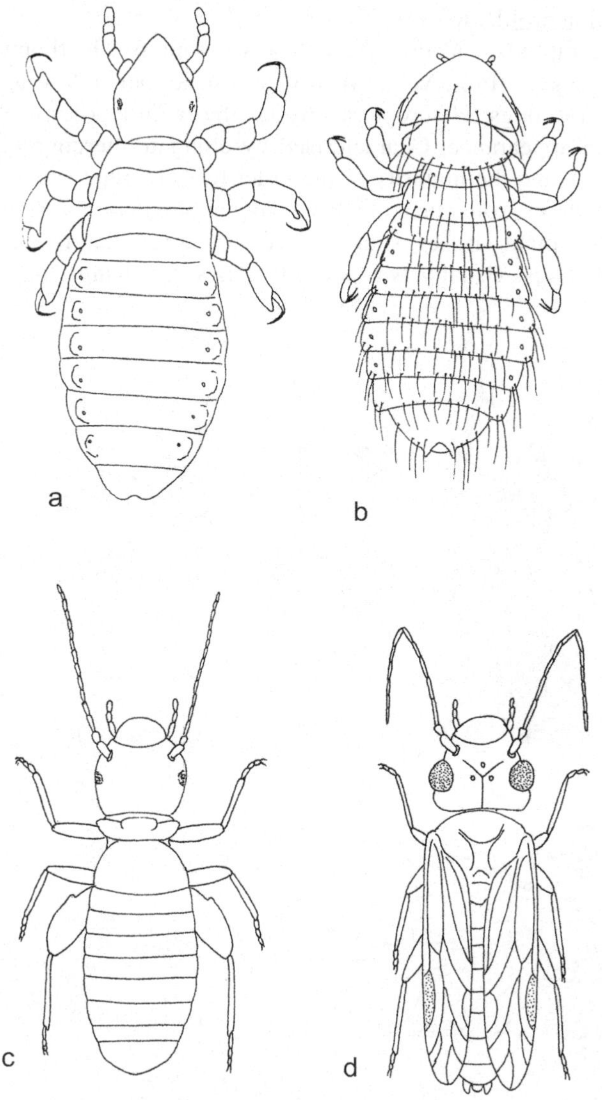

Fig. A9 General appearance of Psocodea: (**a, b**) parasitic lice (Anoplura, Amblycera); (**c, d**) barklice (wingless Liposcelididae, winged psocomorphan)

have either chewing (mammal and bird hosts) or sucking (mammal hosts only) mouthparts. Barklice and booklice, also only a few mm long but not dorsoventrally flattened, are free-living. Many of them live on vegetation or in leaf litter where they graze on microflora, fungi, and organic debris; slender antennae; chewing mouthparts with mandibles asymmetrical and lacinia rod-like; legs slender, tarsi not thickened or hooked; winged or with wings shortened (brachypterous) or absent; fore wing usually larger than hind wing.

(conservation profile: low)

Hemiptera (aphids, cicadas, scale insects, whitefly, leafhoppers and allied groups, true bugs). (Fig. A10). Mostly terrestrial but a few aquatic lineages amongst the true bugs (Heteroptera); by far the largest order in this series, and very varied in appearance. Characterised by elongate piercing suctorial mouthparts ('stylets'), and the majority of the order feed on plant sap as a major group of insect herbivores, some taxa (such as many aphids) are notorious as crop pests or as vectors of plant viruses; some Heteroptera are predators, and include some blood-feeding bugs. Usually two pairs of wings, but wings may be reduced or

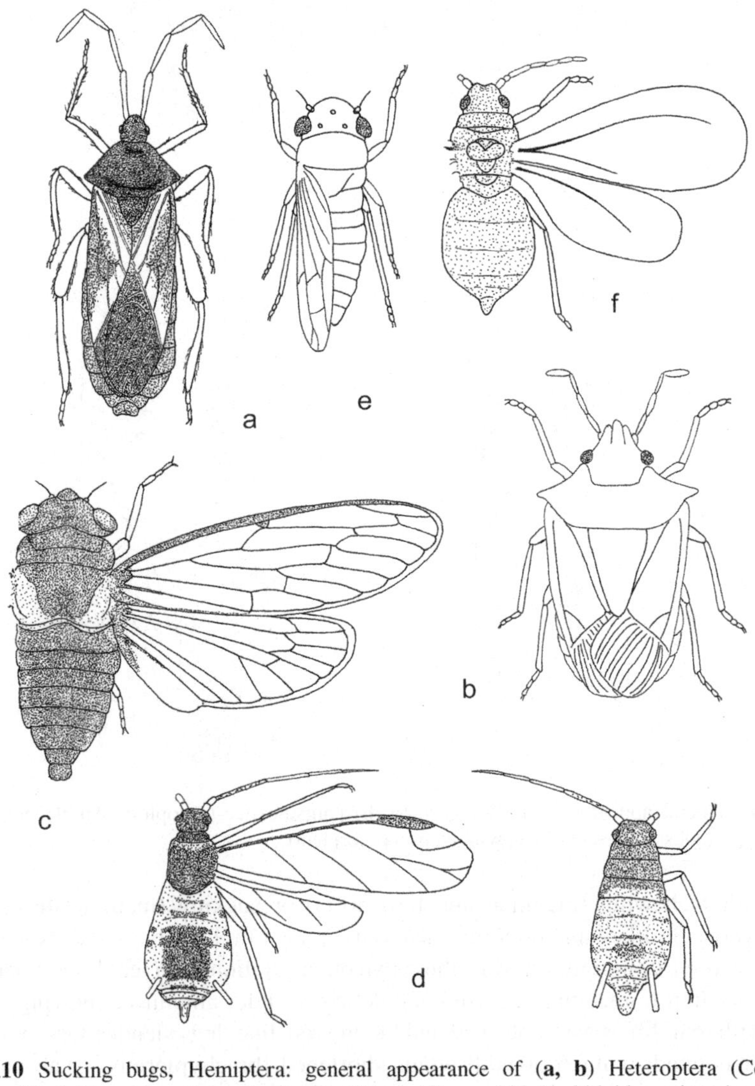

Fig. A10 Sucking bugs, Hemiptera: general appearance of (**a, b**) Heteroptera (Coreidae, Pentatomidae); Homoptera: (**c**) a cicada (Cicadidae), (**d**) an aphid, (Aphidoidea, winged and wingless forms); (**e**) a leafhopper (Cicadellidae); (**f**) a whitefly (Aleyrodidae)

absent. Most groups belong to the suborder Homoptera, in which the fore wing has uniform texture throughout, but the 'true bugs' are distinguished as 'Heteroptera', with the fore wing basally thickened and the apical region more membranous. Small to large.

(conservation profile: low, occasionally higher)

Thysanoptera (thrips). (Fig. A11). Terrestrial. Slender elongate insects, most only a few mm long; winged or wingless; if winged, the wings are long and narrow with only one to three longitudinal veins, and have a marginal fringe of long hairs;

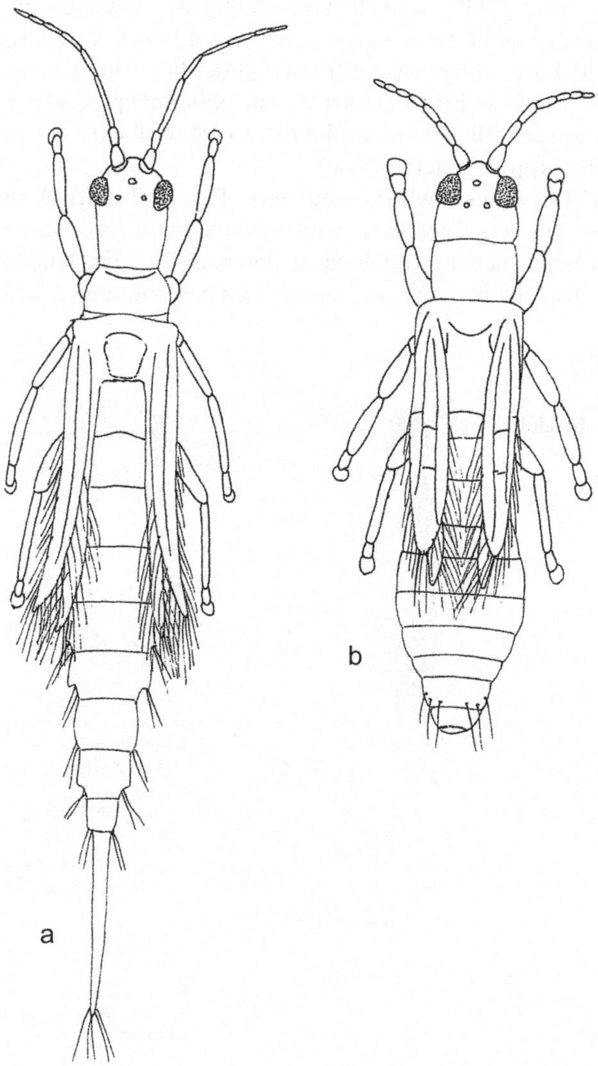

Fig. A11 Thrips: the two major forms of Thysanoptera as (**a**) Tubulifera: Phlaeothripidae and (**b**) Terebrantia: Thripidae

mouthparts asymmetrical, used to abrade food and ingest juices from it; many with apex of abdomen elongate.

(conservation profile: low)

B. The following orders, regarded as the 'most advanced' insects, have a complete metamorphosis, in which the larva and adult are separated by a pupal stage, and may have very different form and lifestyle; wings develop internally and are not visible until the adult stage. These insects are termed 'endopterygote'.

Megaloptera (alderflies, dobsonflies). (Fig. A12 a, b). Aquatic. Regarded as the most primitive endopterygote insects; large, with larvae aquatic and adults usually found close to water. Adults with chewing mouthparts; antennae long and slender; two pairs of broad membranous wings; legs slender, tarsi five-segmented; abdomen soft and flexible. Larvae elongate, with seven (alderflies: Sialidae) or eight (dobsonflies: Corydalidae) pairs of filamentous lateral abdominal processes (gills) and long posterior process (alderflies) or clawed prolegs (dobsonflies).

(conservation profile: generally low)

Neuroptera (lacewings, owlflies, antlions). (Fig. A12 c, d). A diverse order of mostly predatory insects. Most taxa are terrestrial, but a few families have aquatic larvae. Adults with chewing mouthparts; antennae slender, lengths differ across taxa, occasionally 'clubbed' at apex; usually two pairs of large wings with complex

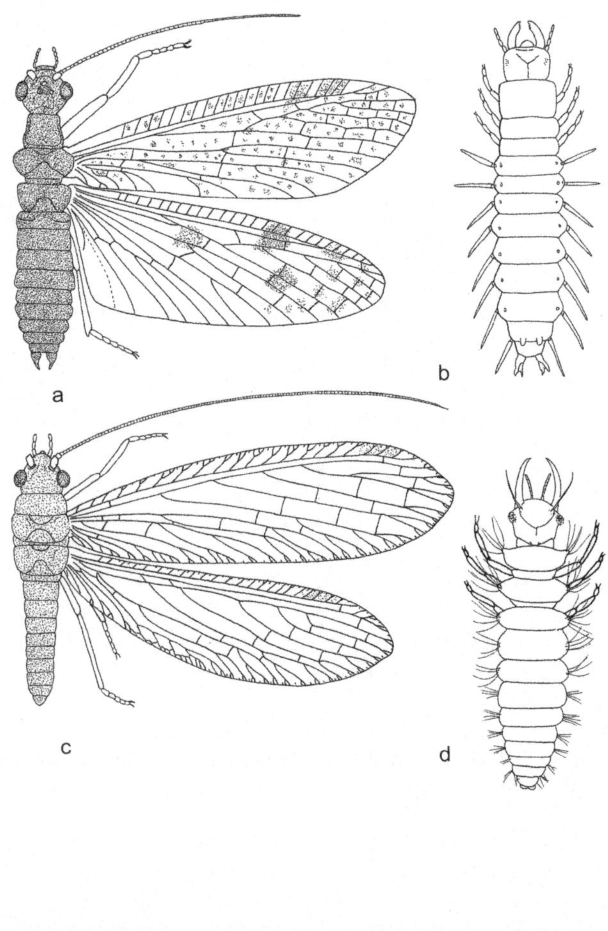

Fig. A12 General appearance of Megaloptera: (**a, b**) adult (*Archichauliodes*) and aquatic larva, and Neuroptera: (**c, d**) adult (*Micromus*), larva of Chrysopidae

venation and held 'tentlike' over the abdomen when not in use. Larvae with suctorial mouthparts, the mandible and maxilla of each side elongated and form a piercing tube for ingesting prey contents; many are active predators on small insects (some valued as predators of pest aphids and others on crops); others (such as many antlions) are ambush predators.

(conservation profile: generally low)

Coleoptera (beetles). (Fig. A13). Often considered the largest insect order and very widespread across terrestrial ecosystems. Mostly terrestrial, but few aquatic families. Ecologically very varied, but characterised by fore wing hardened to form protective 'shell' ('elytron') usually extending over whole/most of abdomen when insect at rest, sometimes very short; hind wing membranous; prothorax large; legs

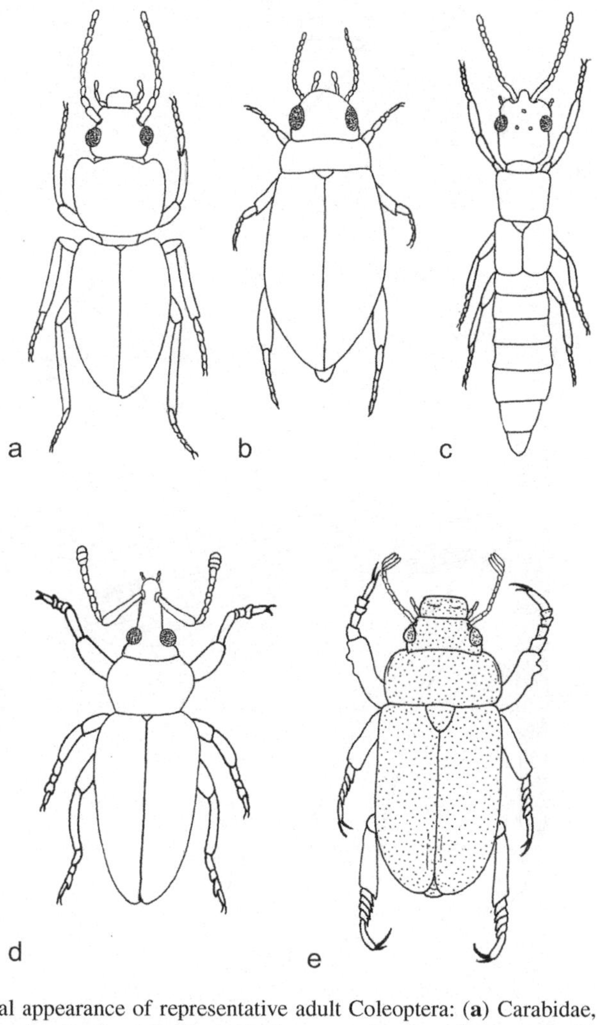

Fig. A13 General appearance of representative adult Coleoptera: (**a**) Carabidae, ground beetle, (**b**) Dytiscidae, water beetle, (**c**) Staphylinidae, rove beetle, (**d**) Curculionidae, weevil, (**e**) Scarabaeidae, Christmas beetle

usually with four- or five-segmented tarsi. Appearance of many of the major groups of beetles is distinctive. Small to large.

(conservation profile: moderate to high, as one of the most popular and ecologically diverse insect orders)

Strepsiptera (twisted-wing insects). (Fig. A14 a). Terrestrial, internal parasites of Hemiptera or social Hymenoptera. Small, males up to about 4 mm long; mouthparts

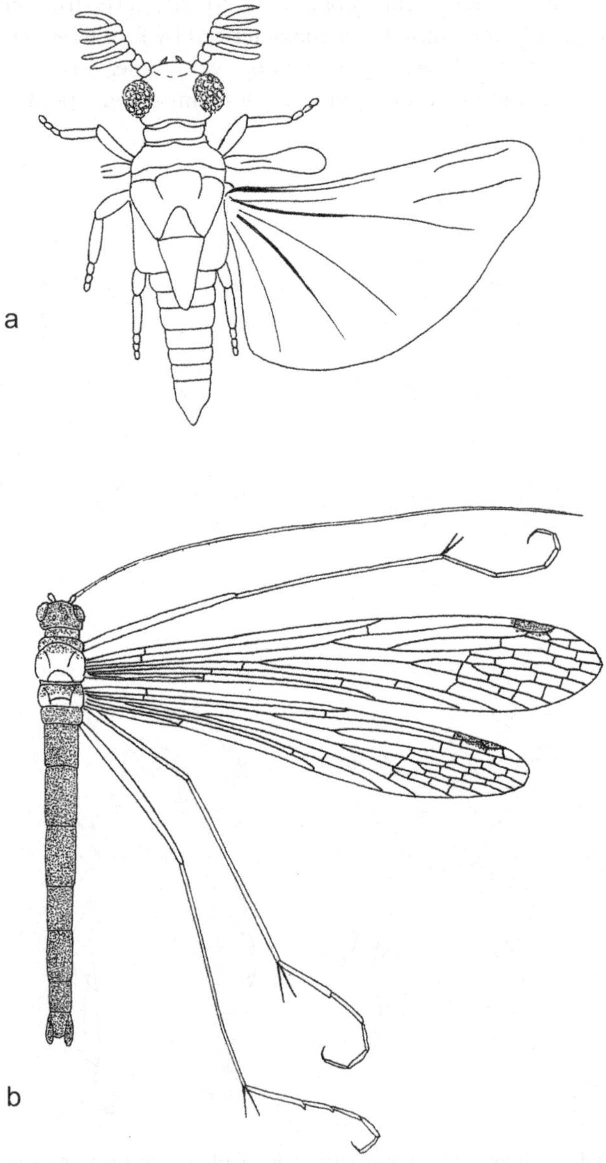

Fig. A14 General appearance of (**a**) Strepsiptera, winged male (Halictophagidae) and (**b**) Mecoptera (Bittacidae)

rudimentary. Strong sexual dimorphism: males with short 'club-like' fore wing and large hind wing, and antennae may be strongly branched; females apterous, 'grublike', found protruding from the host abdomen rather than free-living.

(conservation profile: low)

Mecoptera (scorpionflies, hanging flies). (Fig. A14 b). Terrestrial. Adult: predatory, with head usually elongated into a narrow 'rostrum' with chewing mouthparts at apex; antennae long, slender; usually two pairs of long slender wings; legs long with raptorial five-segmented tarsi; apex of male abdomen in many taxa expanded and curved dorsally to resemble appearance of a 'scorpion tail' (hence the common name). Larvae elongate, with poorly defined abdominal prolegs and groups of eyespots ('stemmata') giving appearance of compound eyes. Most species moderate-sized.

(conservation profile: low)

Diptera (flies, midges, mosquitoes, craneflies, marchflies, and others). (Fig. A15). Mostly terrestrial, but few aquatic families. Characterised by fore wing fully

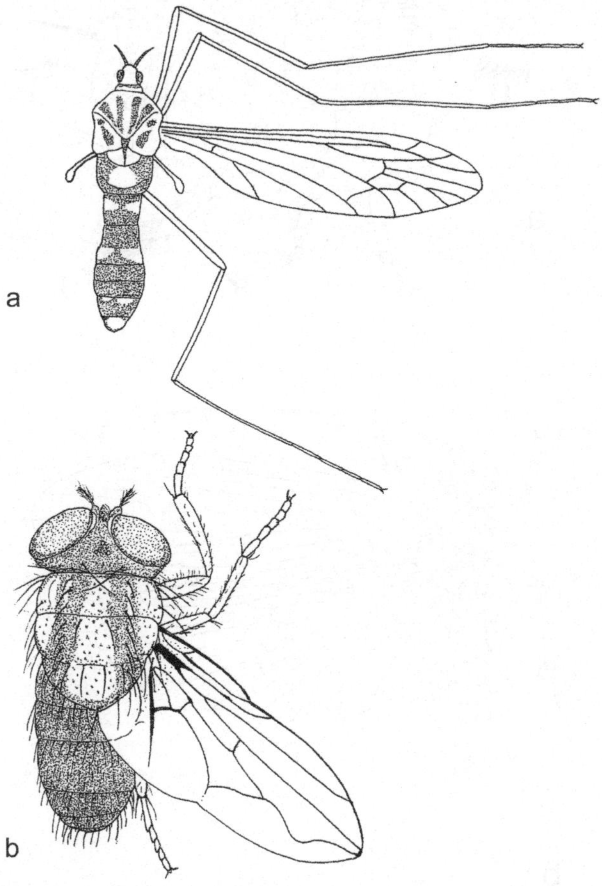

Fig. A15 The two major groups of Diptera: (**a**) Nematocera (Tipulidae), (**b**) Brachycera (Muscidae)

formed, hind wing reduced to small sensory structure ('haltere'); mouthparts suctorial or piercing—in most higher flies as an expanded spongelike labellum that imbibes semiliquid food; antennae often short; mesothorax often large; legs slender, often long. Two major groups differ considerably: (1) the more primitive Nematocera—craneflies, mosquitoes, and their allies—are usually elongate and slender, with relatively long antennae, and larvae with a well-defined head capsule with opposable chewing mandibles; many Nematocera have aquatic larvae; (2) the more advanced Brachycera (typical flies) with very short antennae and larvae ('maggots') with the head capsule reduced or absent and the mandibles hook-shaped and moving vertically. Small to medium, few species larger.

(conservation profile; low to moderate)

Siphonaptera (fleas). (Fig. A16 a). Terrestrial. All are small (a few mm) ectoparasites of mammals or birds. Wingless, body laterally compressed, strong hind legs

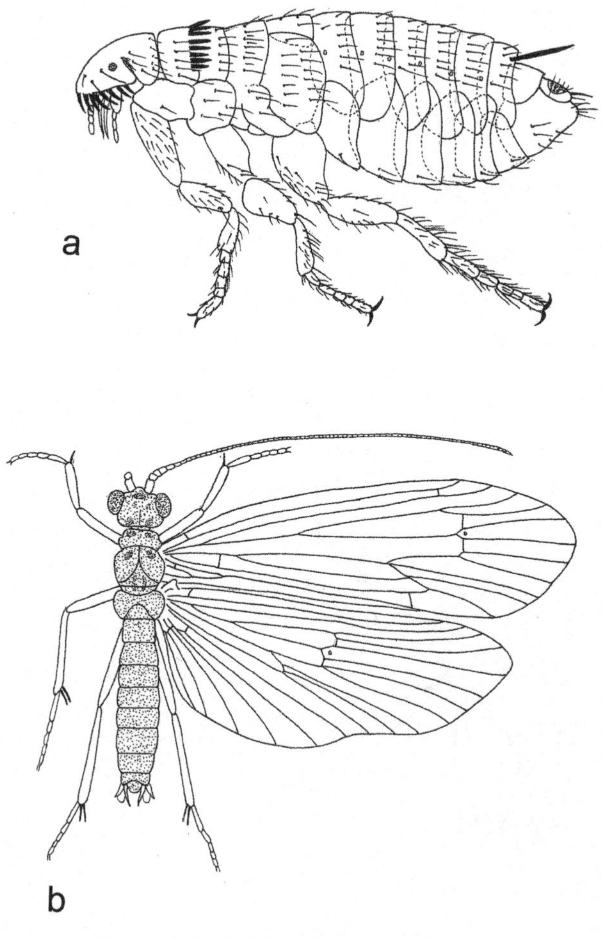

Fig. A16 General appearance of (**a**) a flea, Siphonaptera and (**b**) a caddisfly, Trichoptera

(jumping); compound eyes absent; short three-segmented antennae in groove on side of head; piercing or sucking mouthparts. Larvae elongate, lacking limbs but often with long 'bristles'; chewing mouthparts.

(conservation profile: low)

Trichoptera (caddisflies). (Fig. A16 b). Aquatic, and the major group of freshwater endopterygote insects. Adults with elongate wings, densely covered with short hairs, hind wing often broader than fore wing; chewing mouthparts. Larvae of most taxa form cases of vegetation or mineral material, with their form of taxonomic importance; head and chewing mouthparts well developed; thoracic legs long and slender; many species with filamentous gills on abdomen; posterior pair of anal prolegs. Most species 1–3 cm body length.

(conservation profile: moderate to high, as diverse constituent of indicator series in freshwater environments [EPT])

Lepidoptera (moths, butterflies). (Fig. A17). Almost all terrestrial. Characterised largely by the wings being covered with small flattened scales, often brightly coloured and with their pattern rendering many species individually recognisable, and a coiled suctorial proboscis (except for a few primitive moths with mandibles) which can be extended and used to feed on floral nectar. Larvae (caterpillars) are mainly chewing herbivores, many with fleshy abdominal prolegs and posterior anal 'claspers'. Although often regarded as distinct, and as useful descriptive terms, 'moths' and 'butterflies' are somewhat artificial categories. Small to large.

(conservation profile: overall the highest amongst insects as the most popular and well-studied group, but major gaps in interest persist for most smaller moth species)

Hymenoptera (sawflies, ants, bees, wasps). (Fig. A18). Terrestrial, and biologically very diverse. Chewing mouthparts; many with two pairs of wings, when fore wing larger than hind wing, but many are wingless (e.g. worker ants); venation

Fig. A17 General appearance of representative Lepidoptera: (**a**) a 'moth' (Erebidae: Arctiinae: *Utetheisa* sp.; (**b**) a 'butterfly' (Papilionidae: *Papilio aegeus*, Orchard swallowtail)

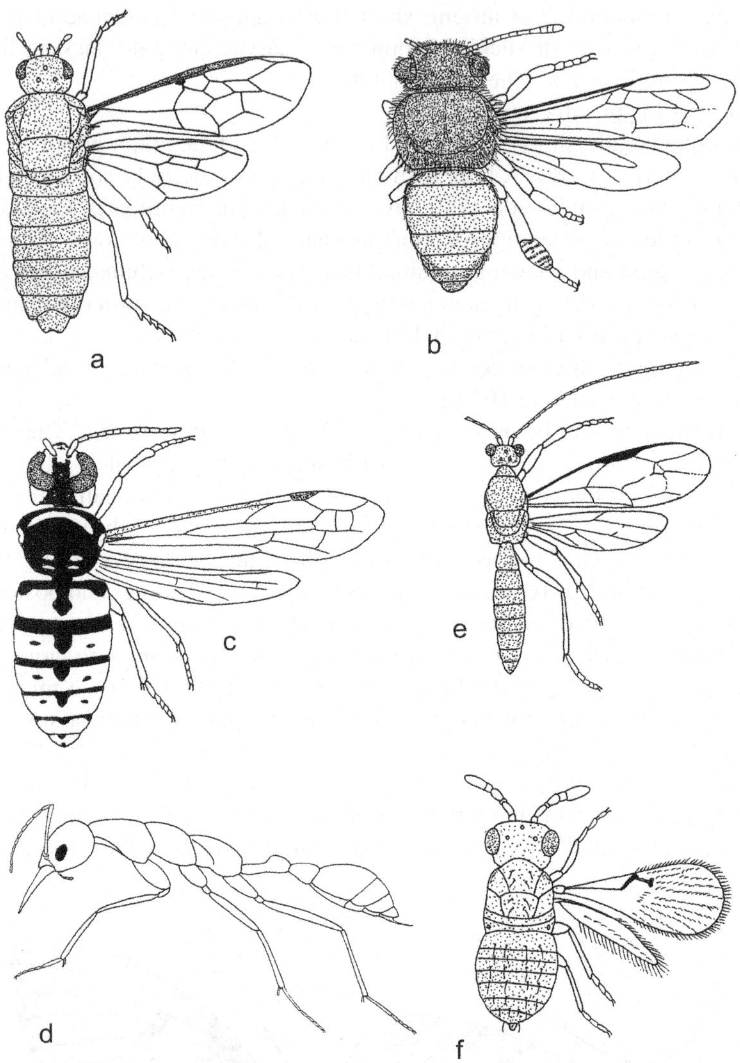

Fig. A18 General appearance of representative Hymenoptera: (**a**) a sawfly (Pergidae), (**b**) a bee (Apidae), (**c**) a wasp (Vespidae), (**d**) an ant (Formicidae, in profile) (**b–d** all social); parasitoid wasps (**e**) Ichneumonoidea: Ichneumonidae, (**f**) Chalcidoidea: Trichogrammatidae, this amongst the smallest wasps

complex to extremely reduced. Two major groups or suborders: (1) Symphyta, the sawflies, regarded as the least specialised Hymenoptera, with no 'waist' and associated mainly with vegetation, as herbivores (larvae of some in Australia are the 'spitfire grubs' on eucalypts), and (2) the Apocrita, all ants, bees, and wasps and by far the larger group, in which there is a narrow constriction (a 'waist') at the second abdominal segment. Apocrita comprises two large groups: (1) the vast array of mostly small wasps that are parasitoids of (usually) other insects, many of them having very specific host requirements, and many are also small, and (2) Aculeata including the stinging ('aculeate') ants, bees, and wasps more familiar to many people. Very small to moderate-sized.

(conservation profile: very uneven. Aculeates, including pollinators such as bees, very high; some ants recognised as important in environmental assessments; most others, particularly parasitoids, generally no to low conservation attention)

References

Baird IRC, Burgin S (2016) Conservation of a groundwater-dependent mire-dwelling dragonfly: implications of multiple threatening processes. J Insect Conserv 20:165–178

Geyle HM, Braby MF, Andren M, Beaver EP, Bell P (et al.) (2021) Butterflies on the brink; identifying the Australian butterflies (Lepidoptera) most at risk of extinction. Austral Entomol 60: 98–110

Glatz RV, Fagan-Jeffries EP, Tetley SP, Austin AD (2022) Enigma moth parasitoid: a new Australian cyclostome genus and species, *Ovaustra aurantia* Tetley, Glatz & Fagan-Jeffries, gen. et sp. nov. (Hymenoptera: Braconidae) from Kangaroo Island. Austral Entomol 61:420–432. https://doi.org/10.1111/aen.12627

Harvey MS, Yen AL (1989) Worms to wasps. An illustrated guide to Australia's terrestrial invertebrates. Oxford University Press, Melbourne

Hawking JH (1999) An outline of the current conservation status of Australian dragonflies. In: Ponder W, Lunney D (eds) The other 99%. The conservation and biodiversity of invertebrates. Trans R Zool Soc NSW, Mosman, pp 354–360

Hill L, Michaelis FB (1988) Conservation of insects and related wildlife. Occasional Paper No. 13. Australian National Parks and Wildlife Service, Canberra

Hogendoorn K, Glatz RV, Leijs R (2021) Conservation management of the green carpenter bee *Xylocopa aerata* (Hymenoptera: Apidae) through provision of artificial nesting substrates. Austral Entomol 60:82–88. https://doi.org/10.1111/aen.12510

Naumann ID (ed) (1991) The insects of Australia (2 vols). Melbourne University Press, Melbourne

New TR (1992) Introductory entomology for Australian students. University of New South Wales Press, Sydney

New TR (1996) Name that insect: a guide to the insects of southeastern Australia. Oxford University Press, Melbourne

New TR (2020) Insect conservation and Australia's inland waters. Springer, Cham

Rentz DCF (1993) Orthopteroid insects in threatened habitats in Australia. In: Gaston KJ, New TR, Samways MJ (eds) Perspectives on insect conservation. Intercept, Andover, pp 125–138

Samways MJ, Simaika JP (2016) Manual of freshwater assessment for South Africa: dragonfly biotic index. Suricata 2,. South African National Biodiversity Institute, Pretoria

Sands DPA, New TR (2002) The action plan for Australian butterflies. Environment Australia, Canberra

Shaw MR (2006) Habitat considerations for parasitic wasps (Hymenoptera). J Insect Conserv 10:117–127

T. R. New, *Insect Conservation in Australia: Why and How*,
https://doi.org/10.1007/978-3-031-66631-5

Shaw MR, Hochberg ME (2001) The neglect of parasitic Hymenoptera in insect conservation strategies: the British fauna as a prime example. J Insect Conserv 5:253–263

Taylor RW (1978) *Nothomyrmecia macrops*: a living-fossil ant rediscovered. Science 201:979–985. https://doi.org/10.1126/science.201.4360.979

Theischinger G, Hawking J, Orr A (2021) The complete field guide to dragonflies of Australia, 2nd edn. CSIRO Publishing, Melbourne

Vorster C, Samways MJ, Simaika JP, Kipping J, Clausnitzer V (et al.) (2020) Development of a new continental-scale index for freshwater assessment based on dragonfly assemblages. Ecol Indic 109: 105819. https://doi.org/10.1016/j.ecolind.2019.105819

Wilkinson R (2014) Return of the phasmid. Australia's rarest insect fights back from the brink of extinction. Media Dynamics, Windsor, Queensland

Index

T. R. New, *Insect Conservation in Australia: Why and How*,
https://doi.org/10.1007/978-3-031-66631-5

SPRINGER NATURE

GPSR Compliance

The European Union's (EU) General Product Safety Regulation (GPSR) is a set of rules that requires consumer products to be safe and our obligations to ensure this.

If you have any concerns about our products, you can contact us on ProductSafety@springernature.com

In case Publisher is established outside the EU, the EU authorized representative is:

Springer Nature Customer Service Center GmbH
Europaplatz 3
69115 Heidelberg, Germany

The manufacturer's authorised representative in the EU is Springer Nature Customer Service Centre GmbH, Europaplatz 3, 69115 Heidelberg, Germany. If you have any concerns regarding our products, please contact ProductSafety@springernature.com

Printed and bound by CPI Group (UK) Ltd, Croydon, CR0 4YY
30/09/2026
02225589-0004